Social Support Networks, Coping and Positive Aging Among the Community-Dwelling Elderly in Hong Kong

Susu Liu

Social Support Networks, Coping and Positive Aging Among the Community-Dwelling Elderly in Hong Kong

華中科技大學出版社
Huazhong University of Science and Technology Press

Springer

Susu Liu
Department of Sociology and Social Work
Soochow University
Suzhou, China

ISBN 978-981-13-3782-6 ISBN 978-981-13-3783-3 (eBook)
https://doi.org/10.1007/978-981-13-3783-3

Jointly published with Huazhong University of Science and Technology Press, Wuhan, China

The printed edition is not for sale in China Mainland. Customers from China Mainland please order the print book from Huazhong University of Science and Technology Press.
ISBN of the China Mainland edition: 978-7-5680-3825-6

Library of Congress Control Number: 2018965425

This Springer imprint is published by the registered company Springer Nature Singapore Pte Ltd.
The registered company address is: 152 Beach Road, #21-01/04 Gateway East, Singapore 189721, Singapore

Contents

List of Figures

List of Tables

Abstract

Fueled by years of the lowest mortality and fertility rates worldwide, Hong Kong has become one of the fastest-aging societies in the world with a high proportion of older adults among Asia Pacific regions. Against this background, the aging situation has elicited new risks and opportunities to facilitate a positive life for the older adults. In light of the aging situation, attention is drawn to the quality of aging, such that the concepts of *healthy aging, active aging, successful aging*, and *positive aging* are frequently encountered. Positive aging was introduced in this study as a pressing concept for longer-life societies, which answers the question that how older people could positively live their longer lives to the full. Specifically, the present study examined the contribution of social support networks (e.g., structure, function, and emotional closeness) and coping (e.g., problem, emotion, and avoidance) to positive aging inspired by a convoy approach.

Grounded in a holistic review of theories and literature on the research topic, the present study constructed a conceptual framework and proposed four main research questions and 38 hypotheses. A mixed-method approach was adopted to explicate the situation of positive aging and the contribution of social support networks and coping with a particular interest in family and peer support networks.

The present research consisted of two phases of studies. The pilot study was a qualitative study that employed in-depth interview with 16 older participants from two District Elderly Community Center (DECC) in Hong Kong. Interviewees were asked to talk about their opinions towards the main research questions in the constructed interview guidelines. Thematic findings were collected and analyzed, from which inspirations of scale construction in the main study were drawn. The main study was a cross-sectional survey of 393 older people who were conveniently selected from six DECC in Hong Kong through a multi-stage sampling method. Face-to-face interviews were carried out by trained student interviewers using a constructed questionnaire.

The present study depicted an indigenous landscape of positive aging of older Hong Kong Chinese in the community center with lower-middle socioeconomic statuses. It clearly presented the contributing factors including perceived socioeconomic status, social network size, family/peer social networks, emotional closeness, and problem/emotion coping. The results of the study indicated that family and peer support constituted the mainstay of support networks of the elderly, of which family and peer emotional closeness were positively predictive of positive aging. Peer groups, including center members, friends, old colleagues, neighbors, and old classmates, can provide vital emotional support and companionship support to older people in the community, assisting them in coping with stressors. Moreover, the study found that a larger close social network might not be equivalent to higher quality of close relationships. It is the depth of emotional closeness, namely, close interaction and intimacy with social partners that makes more contributions to positive aging. Among the three coping strategies (e.g., problem, emotion, and avoidance coping), avoidance coping was prevalently used by older adults in all age groups. Participants with higher education background tended to use more problem and emotion coping. Additionally, problem and emotion coping were found to mediate the relationship between social support network and positive aging. With regard to the role of domestic helper in providing social support to the elderly, the instrumental and companionship support provided by domestic helpers were beneficial. There is potential in bringing more domestic helpers in elderly care and improving the service quality such that the goal of *aging in place* (AiP) can be promoted in Hong Kong.

Significant findings of this study are essential to facilitate the Hong Kong elderly to age well in the community, shedding light on the meanings and benefits of the notion AiP. The study echoes a few implications and contributions to policy makers (e.g., facilitate peer support networks and build dynamic elderly community center), social services (e.g., develop more center members), and common families (e.g., enhance positive interaction within generations and support caring and productive engagement) to assist the elderly to lead a positive life in the community. On the practice level, it is suggested that the community and families should pay more attention to the physiologic, psychological, and societal needs of older adults, mobilize resources, and develop emotional closeness together with the elderly.

The limitations of this study have been duly noted. On the academic level, longitudinal studies are recommended to unravel the unsolved research questions from a lifespan perspective. In addition, future research is suggested to enrich the framework and predictors of the convoy model. In the meantime, the results and implications of the current research regarding the concept, framework, and measurement of positive aging require further demonstration and extension in other populations at home and abroad, motivated by the desire to promote well-being in later life.

Chapter 1
Introduction

Hong Kong, as a special administrative region of the People's Republic of China (PRC), embraces a large density of the Chinese population, of which the older population is booming. It is one of the fastest-aging societies in the world with a high proportion of older adults among the Asia Pacific regions. Though many older adults are healthy, the burgeoning older population has elicited new risks and opportunities to facilitate a positive experience of aging for older adults. In Hong Kong, the number of community-dwelling elderly accounts for over 90% of the population of older adults. As to the older adults living in the community, how can they live positively and continue to contribute to family and society at an old age? The present study aims to examine the concept of positive aging, identifying its definition, components and implications, and exploring two main influencing variables (e.g., social support networks and coping), taking Hong Kong as a case study.

This chapter is to set the stage for the following study and consists of five parts. Part One describes the background in regard to the aging situation of the world and Hong Kong, the challenges and opportunities brought by an aging society and the policy directions for the elderly in Hong Kong. Part Two states the research problems examined in the present study. Part Three introduces the research rationale and significance of the study. Part Four presents the expected research objectives of the study. Lastly, the layout of the thesis is displayed to give a holistic view for readers in Part Five.

1.1 Background

1.1.1 The Situation of Global Aging

According to the definition of the UN, if the population of 60-year-old accounts for 10% of the total, or the population of those over 65 reaches 7% in a country or

S. Liu, *Social Support Networks, Coping and Positive Aging Among the Community-Dwelling Elderly in Hong Kong*,
https://doi.org/10.1007/978-981-13-3783-3_1

region, we can say the country or region has become a society of an aging population (Population Division of the United Nations, 2002). Indeed, it has become common knowledge that the world has stepped into a new era of societies with increasing aging populations. According to the statistics, since 1950, the proportion of older people has been rising steadily, ranging from 8% in 1950 to 11% in 2009, and is predicted to reach 22% in 2050 (World Population Aging, 2009). Furthermore, as the proportion of the world's older population continues to increase, a series of aging issues are gradually attracting more attention from a global perspective. In the latest report of global aging, it is proclaimed that the whole world is experiencing an unprecedented trend of population aging, which is a process without parallel in the history of humanity (Word Population Aging, 2009). Additionally, the profound impacts brought by an aging population might produce potential major consequences and various implications in all facets of life, including the economic area, social sphere, and political arena, etc. Against this background, the lower fertility rate and continuous reduction of adult mortality could account for the main reasons why we are undergoing this unprecedented demographic change in recent years (World Health Organization [WHO], 2008). In light of the specific phenomenon, a number of far-reaching economic and societal adjustments should be implemented to face the challenges and opportunities accompanied by population aging.

Although this irresistible tide has expanded the portion of older adults among the total population and caused new risks in elderly care and other social welfare affairs, potential opportunities are there to be grasped, to construct a healthy, positive and contributive aging society (Chi, Chappell, & Lubben, 2001; Leung, Chen, Lue, & Hsu, 2007). As to the older persons living in the new twenty-first century, their roles are not regarded as inflexible as before because new notions have brought in new meanings to their later life. Consequently, an older adult can still choose a reasonable and pleasing lifestyle in spite of the normal decline in biological and physical functions (Walker, 2002). Furthermore, scholars in academia also attempt to facilitate the spread of gerontological concepts such as successful aging, positive aging, active aging, and so on, aiming to enhance the quality of life among older people, which, therefore, alleviates the pressures brought by fast aging (Bowling & Dieppe, 2005; Gergen & Gergen, 2001; Nagchoudhuri, McBride, Thirupathy, Morrow-Howell, & Tang, 2005; Walker, 2002). Hence, this background poses an interesting question for the aging society—how to get through the challenges in later life by virtue of the available resources from a positive perspective and continue contributing to families and societies?

1.1.2 Hong Kong: A Fast Aging Society

Hong Kong is one of the fastest-aging societies in the world. It has been a Special Administrative Region of the PRC since 1997. Although it has been a British colony for over 150 years, most of the indigenous citizens are predominately Chinese. In comparison to the average pace of growing old, Hong Kong has a comparatively accelerated rate of aging, doubling its proportion of older people from 6.6 to 13.9% in the last 30 years (Cheng, Lum, Lam, & Fung, 2013). By 2031, one in every four people is expected to be aged 65 years or older (Census and Statistics Department, 2010). At present, the decreased fertility rate (1.04 per female) (The World Bank, 2010) and increased life expectancy (79.7 years for males and 85.9 years for females) (Census and Statistics Department, 2010) has given rise to the prominent issue of aging. Nevertheless, the universal aging phenomenon should not be solely regarded as a negative problem, and appropriate measures could and should be taken to develop a healthy and positive aging society (Leung, Chen, Lue, & Hsu, 2007).

Dating back to older times, the Hong Kong government recognized the pressing challenges of taking care of the elderly in an aging society, and promulgated the *Hong Kong White Paper on Social Welfare* in 1979 (Gui & Huang, 2003). Although government intervention in social welfare was believed to be necessary only for people who were most deprived, or people who were lacking family support, the government has become more proactive after returning to the PRC in 1997 (Cheng et al., 2013). Improvements in welfare and service programs, especially for the aged, could be seen. Since then, services for the elderly have experienced a speedy development and gradually generated voluminous multi-level services. At present, in order to achieve the goal *A Sense of Security, A Sense of Belonging and A Feeling of Worthiness for the Elderly*, which was first put forward in 1998, Hong Kong has established a relatively stable system of elderly services (Hong Kong Policy Research Institute, 2003). The integrated elderly service system encompasses five prime categories: community support services, elderly institutional services, medical and health care services, an elderly housing service, and financial aids. Hong Kong is getting down to launching a Long-Term Care System for the purpose of enabling the frail elderly to live with dignity, and maintain an optimal level of quality of life (Chi, 2001). It mainly contains community care services, residential services and various corresponding subservices for older adults with different needs.

1.1.3 Positive Aging: A Productive Approach to Well-Being in Old Age

Scholars have endeavored to discover the theoretical and practical approaches to facilitate the well-being of older adults. In social gerontology, positive aging and other relevant concepts including successful aging, active aging, and productive aging are currently regarded as a prevalent approach to describe and examine older people's later life (Fernandez-Ballesteros, 2011; Reichstadt, Depp, Palinkas, Folsom, & Jeste, 2007; Tan, Ward, & Ziaian, 2011). Some social gerontologists argue that these terms are largely similar in their definitions, implications and conceptual frameworks, which all contain factors that depict well-being in old age (Bowling, 1993; Bowling & Iliffe, 2006; Rowe & Kahn, 1997). The concept of positive aging may refer to life satisfaction, coping behaviors, and physical ability, which includes the retention of social support networks associated with well-being (Bowling, 1993). Though it is accepted that positive aging is endowed with numerous similar meanings which are derived from the concept of successful aging (Bowling, 1993), this study will take positive aging as the research focus and endeavor to reconstruct its conceptual framework with reference to the Chinese elderly.

Positive aging, as a newly proposed concept originated from Western countries, has received much interest from academic research across the world (Rowe & Kahn, 1997; Bowling, 1993; Ryff, 1982). However, empirical research on positive aging among Chinese citizens is still sparse. Given its existing significance of achieving a healthy, positive and productive later life, the present study takes the concept of positive aging as the dependent variable and attempts to examine its influencing factors such as social support networks, coping strategies, emotional closeness and demographic variables among older Hong Kong Chinese.

1.1.4 Hong Kong Elderly: Living Under a Blending Chinese Culture

It has long been recognized that Hong Kong has its unique features in many aspects as a region where the East meets the West. Although Hong Kong society had been immersed in colonial culture for over 150 years and witnessed substantial dramatic social and economic changes, yet the traditional Chinese culture (e.g., Confucianism, family self-reliance ideology) has been valued and embraced by the majority of Chinese citizens, especially the elders. Therefore, it is worthy to notice that Hong Kong citizens' endorsement of culture and value is more complicated than that in mainland China. Given the prolonged exposure to infiltration capitalist ideology under colonial rule and the gradually increasing communication with the mainstream culture of traditional Chinese, it is significant to explore the cultural impacts on older Hong Kong Chinese.

However, like a few metropolises in mainland China, Hong Kong is experiencing an era in which traditional Chinese culture is being eroded but still prevails. At present, it is more acceptable that the individualism trend in Hong Kong is becoming more popular, which might create a number of pressures in terms of family life and elderly caregiving (Chan, 2009). In regard to the blending of cultural ways with which older people are now indoctrinated, would they still be able to enjoy traditional Confucian values in their later life? Do they have their unique thinking and lifestyles in terms of aging well in Hong Kong? Thus, the unique cultural factors existing in Hong Kong society might have an influence on older people's positive aging, which calls for in-depth exploration in this study.

1.2 Research Gap and Research Problems

In regard to the evidence of previous literature and interests of research, the present study puts forward the research problems after identifying the existing research gap.

1.2.1 What Are the Components of Positive Aging Among the Community-Dwelling Elderly in Contemporary Hong Kong?

The statistic shows that over 90% of the elderly live in the community (Chi, 2001). Compared with older adults living in residential care homes, the community-dwelling elderly have more resources to retain their roles in families and societies. For over 20 years, the objective of elderly services in Hong Kong has been to promote the well-being of the elderly through care in the community and by the community (Hong Kong Government, 1997). Furthermore, the Hong Kong government raised a policy titled *aging in place* (AiP), which was defined as enabling older people to remain living at home and to forestall premature or inappropriate institutionalization (Moore, 2000). The principle of AiP emphasizes the need to allow older individuals to remain living in the locality with which they are familiar for as long as they wish. This principle aims to avoid the risk of older people losing their sense of security when they are faced with removal from a familiar physical and social environment. Although it seems that AiP is more concerned about the frail elderly or the elders with chronic diseases who are in need of being cared for, the agenda embraces several implications such as enhancing the quality of life of older adults in the community and creating a good communal environment for the elders.

Positive aging has been examined in western and Asian countries since decades ago (Bowling, 1993; Fernandez-Ballesteros, 2011; Reichstadt et al, 2007; Chong, Ng, Woo, & Kwan, 2006; Ng et al., 2011). Previous research lent credence to building up the theoretical framework of positive aging, however, to the best of my knowledge, no framework exclusively applied to the community-dwelling elderly in Hong Kong has been formulated and validated. Based on the empirical evidence of positive aging discovered in Hong Kong (Chong et al., 2006; Ng et al., 2011), the researcher intends to identify the main components of positive aging and to depict the image of positive aging life among the community-dwelling elderly who are members of District Elderly Community Center (DECC) in Hong Kong, with the inspirations of AiP. Specifically, the study aims at examining the definition, development and components of the concept of positive aging through a qualitative study and quantitative study.

1.2.2 *What Is the Effect of Social Support Networks and Coping on Positive Aging?*

A burgeoning number of studies concerning positive aging and other psychosocial facets in later life have been conducted to uncover the relationship between them. Derived from the existing research findings, personal characteristics including gender (Depp & Jeste, 2006), socio-economic status (SES, Western, 2002), healthy life styles (Schnittker & John, 2007), receiving social support (Montross et al., 2006) and religious beliefs (Crowther, Parker, Achenbaum, Larimore, & Koenig, 2002) are posited to influence positive aging. Among the psychosocial factors, previous studies have explored the structures of social relationships or social networks (Freeman, 2004; Hill & Dunbar, 2003), the functions of social support (Chi & Chou, 2001), and emotionally close networks (Lang & Carstensen, 1994; Fung, Stoeber, Yeung, & Lang, 2008).

Additionally, as documented in existing literature related to positive aging or successful aging, adaptive approaches (e.g., coping) to balance the gains and losses appearing in later life are widely emphasized (Balts & Balts, 1990; Lang & Carstensen, 1994). Coping strategies are originally defined as effective attempts to manage specific situational demands which are appraised as taxing or exceeding one's ability to adapt (Lazarus & Folkman, 1984). Specifically, coping strategies have been demonstrated to play an important role, enabling the elderly to lead a healthy and positive life in their sunset years (Folkman, Lazarus, Pimley, & Novacek, 1987; Jonker, Comijs, Knipscheer, & Deeg, 2009). Moreover, coping has been demonstrated as a mediator in different scenarios of stress process models (Connor-Smith & Compas, 2002). Evidence has showed that coping mediated the relationship between acculturative stress and psychological stress (Vasquez, 2010). Would coping also affect the relationship between social support networks and positive aging?

However, there is a dearth of empirical studies in this area. Drawing on the above literature review, the present study mainly aims at examining the influence of social support networks and coping on positive aging, taking Hong Kong as the case study.

1.2.3 *What Is the Effect of Emotional Closeness on Positive Aging?*

In the research on social relations or social networks, the convoy of social networks is a model first proposed by Kahn and Antonucci (1980), describing a set of roles providing supportive interactions which are distributed into three layers across one's lifespan. Among the social relations of an individual, persons to whom the individual feels so close that life cannot be imagined without them are placed in the inner circle. The number of close social partners and the degree of closeness (e.g., contact frequency, intimacy) with them constitute the breadth and depth of emotional closeness—one of the important variables that are researched in this study.

Emotional closeness between two persons leads them to want to help each other, especially when the other is in need (Korchmaros & Kenny, 2001). So far, relevant theories and research suggest that close relationships and social support are essential to the mental health of adults (Rathus & O'Leary, 1997). It is also found that strong emotional connections in later life predicted happiness and adjustment (Antonucci & Jakson, 1987). As to older adults, family members and peer groups (e.g., community center members, friends, neighbors, old colleagues, ex-classmates) are the two major sources of emotionally close networks. However, empirical studies in terms of emotional closeness in older Chinese are lacking. Consequently, the present study will explore emotional closeness in later life and its association with positive aging. Do the community-dwelling older adults have a preference for family emotional closeness or for peer emotional closeness? Do both types of emotional closeness positively predict positive aging?

1.2.4 *What Is the Role of Domestic Helpers in Providing Social Support to the Elderly?*

With the development of home-based or community-based care, more and more elderly people choose to stay in their familiar living environment to spend their sunset years. For the reason that adult children are usually engaged with their own career and family, some new caregivers including domestic helpers (DHs), nurses, and home helpers, etc. emerge to provide informal care to older people (Lai, 2009). As to the composition of domestic helpers, most of them are live-in foreign women, followed by a small number of local full-time or part-time helpers. To be able to AiP, older people with deteriorating physical and/or cognitive health usually rely on a combination of formal and informal care including family care. Hong Kong has a unique advantage over many industrialized societies in that many families could employ, at relatively affordable fees, foreign domestic helpers (FDHs) who are mainly women coming from south-east Asian countries to assist in caregiving and to take up household chores. The number of FDHs reached 292,473 by 30 June 2011, reflecting a substantial increase (30.92%) compared with 223,394 in 2005 (Legislative Council, 2011).

Empirical research shows that FDHs are usually responsible for hands-on personal care and household chores (Lan, 2002). Support from DHs was found to be beneficial to those suffering from diseases and disabilities (Lam, Lee, Chung, Lau, Woo, & Kwok, 2010). They might accompany their patients on walks, during doctor visits, or on trips to day care centers. DHs could also represent a commodification of the caring role. A study of eight Chinese dual-earner immigrant families in California found that hiring a FDH to care for an older parent was interpreted as a form of subcontracting filial care (Lan, 2002). A recent study found that frail elders with cognitive or functional impairments had a significantly higher risk of institutionalization both in Hong Kong and in other societies, and that the employment of a DH, usually a non-local one, predicted reduced institutionalization. FDH, therefore, could serve as a protective factor against institutionalization (Chau et al., 2010; Chong, Kwan, Chi, Lou, & Leung, in press). Yet, a perplexing phenomenon is that Hong Kong still has an unusually high institutionalization rate of 6.8% (Chui et al., 2009). Moreover, there is a long waiting time for residential care which is 36 months and 33 months respectively for the subsidized care-and-attention home and nursing home respectively (as at 31 March 2014) (Social Welfare Department [SWD], 2014).

In line with the current situation of employing DHs in elderly care, a couple of research questions are therefore put forward. What is the role of DHs as caregivers to older adults? What kind of support do they provide? What are the pros and cons of employing a DH from the perspective of an older person?

1.3 Rationale and Significance of the Study

A great number of academic studies in the domains of social work, sociology, and psychology have focused on aging issues. Gerontologists incline to lay their sight on the living conditions, health status and well-being related to older adults in the past few years (Cheng, Lee, Chan, Leung, & Lee, 2009; Lau, Chi, & McKenna, 1998; Lazarus & DeLongis, 1983; Okun & Stock, 1987). However, few empirical studies have examined the well-being issue of how older adults could lead a healthy, positive and productive later life with personal resources and external social resources in Chinese society. In addition, previous literature has recorded that both social support networks and coping can contribute to well-being in later life; however, the question of whether a specific property of social support networks could trigger a type of coping strategy in affecting positive aging is rarely examined. Therefore, the current research aims to fill this research gap by exploring the components of positive aging and identifying the interrelationships between social support networks and coping, as well as their predicted effect on positive aging. Consequently, the current research will conduct a qualitative study (in-depth interview) and a quantitative study (survey) concerning the community-dwelling older Chinese in Hong Kong in order to elucidate the research questions.

The key concept to be explored in this study is positive aging. As an integrated concept depicting an all-sided status of older people in social gerontology, positive aging has been used in past studies to examine the well-being of the elderly in different regions (Bowling, 1993; Chong, Ng, Woo, & Kwan, 2006; Gergen & Gergen, 2001; Woo et al., 2008). Normally this term represents an umbrella notion or a guiding framework to describe the challenges and opportunities that can be encountered in later life, and the role of individual choice in directing how one approaches growing old in this modern and life-extended era (Hill, 2005). Like other similar concepts such as successful aging, active aging and so on, positive aging contains certain components which encompass the general needs and facets of older people's life. However, the essence of positive aging still depends on the specific construct towards this concept when it was applied in different contexts (Bowling, 1993). In this study, a revised conceptual framework of positive aging built on the model proposed by Ng et al. (2011) will be employed to examine whether older citizens are leading a healthy, positive and productive sunset life in Hong Kong. My specialized implication would be posted on the construct of positive aging and the validation of the measure, taking consideration of the characteristics of Chinese elderly living in the community.

One of the research focuses is the concept of a social support network. It is derived from the initial discussions on social support theory in the 1970s (House, 1981; Vaux, 1988). Over the past decades, social support and social networks have been widely and deeply explored among different age groups and social contexts (Wenger, 1984; Wong, 2008; Yeung & Fung, 2007). A social support network is generally adopted to describe the kinds of support that one receives from others, which could constitute a systematic network. Although numerous studies were conducted to explore the social network or social support in later life, few empirical studies were found to examine both the structural and functional properties of social support networks among the Chinese elderly. Older adults, as a relatively vulnerable group with a tendency to experience physical, functional or mental decline, are prone to encounter stressors. Therefore, more concern and support are needed to assist them in getting through the adverse effects occurring in later life (Schwarzer & Leppin, 1991). Furthermore, a few existing empirical and theoretical studies on elders' social support networks in Western countries could provide a reference for the current research. Thus, the present study is expected to have a holistic view of social support networks in Chinese societies such as Hong Kong, especially among the community-dwelling elderly.

As to coping, given its importance in the psychological adjustment of individuals, there is a dearth of empirical studies examining the coping strategies applied in old age. Drawing on the observations in previous sections, the present study endeavors to examine coping among the elderly, to identify the prevalent coping strategies adopted by the elderly and to find out the relation between the social network and positive aging.

In view of the statements concerning the research problems and rationale, the present study embraces four major aspects of significance. First, the present study will propose a revised conceptual framework of positive aging among Chinese elderly, based on the research findings of the qualitative study and inspirations of theoretical models. In the meantime, it will discover the contributing factors that influence positive aging. Second, this study will provide timely, systematic and in-depth information in regard to the structural and functional properties of social support networks, prevalent types of coping strategies and selected demographic variables among the community-dwelling elderly in Hong Kong. Third, the study will generate rich statistical information on the different trajectories between social support networks, coping and positive aging, and propose appropriate recommendations for policy making, social services and research areas in the future. Last but not least, the study expects to identify the role of DHs as support providers to the elderly and to give advice on improving the quality of service.

1.4 Research Objectives

Based on the preliminary reviews, the present study formulates six research objectives: (1) To construct the conceptual framework of positive aging in the Chinese context; (2) To explore the structural (e.g., type, size) and functional properties of social support networks (e.g., instrumental, emotional, companionship and informational support) of the elderly living in the community; (3) To examine coping in later life and find out the prevalent coping strategies; (4) To understand emotional closeness composed of family members and peer groups in later life; (5) To discover how the properties of social support networks, coping and demographic factors affect positive aging; and (6) To identify the role of DHs in providing social support to the elderly.

1.5 Layout of the Thesis

The thesis consists of eleven chapters. This chapter (Chapter One) provides the background of the study, the rationale for the choice of topic, the objectives and significance of the study. In Chap. 2, literature on positive aging, social support network, coping and DHs in elder care is reviewed. Additionally, relevant social gerontological concepts on well-being are also presented and discussed. Chapter 3 first introduces the theoretical background of the study and then outlines the conceptual framework, followed by a list of the research hypotheses. In Chap. 4, the research design is outlined, followed by the introduction of the methodology, which consists of the qualitative study (in-depth interview) and the quantitative study (survey). Chapter 5 presents the thematic findings and implications of the in-depth interview. In Chap. 6, descriptive analysis of the survey, including the respondents' profiles, descriptive statistics of main variables, and correlations between main variables is conducted. Chapter 7 displays the results of hypotheses testing of the survey. The findings of DHs as support providers and discussion on the results are presented in Chap. 8. Chapter 9 discusses the implications and meanings of the major observations and findings of the study. Moreover, the findings that are consistent or inconsistent with existing literature will be identified. Chapter 10 elaborates the implications of the study, including the recommendations on ways to promote positive aging and supportive resources in old age. Additionally, limitations of the study and future research directions will also be discussed. In Chap. 11, the contributions of the research are elucidated. Then the conclusions of the whole study are drawn at last.

References

Antonucci, T. C., & Jackson, J. (1987). Social support, interpersonal efficacy, and health: A life course perspective. In L. L. Carstensen & B. A. Edelstein (Eds.), *Handbook of clinical gerontology* (pp. 291–311). Elmsford, NY: Pergamon Press.

Baltes, P. B., & Baltes, M. M. (1990). Psychological perspectives on successful aging: The model of selective optimization with compensation. In P. B. Baltes & M. M. Baltes (Eds.), *Successful Aging: Perspectives from the behavioral sciences* (pp. 1–34). New York: Cambridge University.

Bowling, A. (1993). The concepts of successful and positive aging. *Family Practice, 10*(4), 449–453.

Bowling, A., & Dieppe, P. (2005). What is successful ageing and who should define it? *British Medical Journal, 331*(7531), 1548–1551.

Bowling, A., & Iliffe, S. (2006). Which model of successful ageing should be used? Baseline findings from a British longitudinal survey of ageing. *Age and Ageing, 35*(6), 607–614.

Census and Statistics Department. (2010). *Statistical tables: Population and Life Events.* Retrieved October 20, 2011 from http://www.censtatd.gov.hk/FileManager/SC/Content_803/population.pdf.

Chan, R. K. H. (2009). Risk, individualization and family: Managing the family in Hong Kong. *Journal of Asian Public Policy, 2*(3), 354–367.

Chau, P. H., Kwok, T., Woo, J., Chan, F., Hui, E., & Chan, K. C. (2010). Disagreement in preference for residential care between family caregivers and elders is greater among cognitively impaired elders group than cognitively intact elders group. *International Journal of Geriatric Psychiatry, 25*(1), 46–54.

Cheng, S. T., Lee, C. K. L., Chan, A., Leung, E. M. F., & Lee, J. J. (2009). Social network types and subjective well-being in Chinese older adults. *The Journals of Gerontology, Series B: Psychological Sciences and Social Sciences, 64*(6), 713–722.

Cheng, S. T., Lum, T., Lam, L. C., & Fung, H. H. (2013). Hong Kong: Embracing a fast aging society with limited welfare. *The Gerontologist, 53*(4), 527–533.

Chi, I. (2001). Long-term care policy for elders in Hong Kong. *Journal of Aging & Social Policy, 13*(2/3), 137–153.

Chi, I., Chappell, N. L., & Lubben, J. E. (2001). *Elderly Chinese in Pacific Rim Countries: Social support and integration.* Hong Kong: Hong Kong University Press.

Chong, A. M. L., Ng, S. H., Woo, J., & Kwan, A. Y. H. (2006). Positive ageing: The views of middle-aged and older adults in Hong Kong. *Ageing & Society, 26*(2), 243–265.

Chui, E. W. T., Ko, S. F. L., Law, C. K. S., Lou, W. Q. V., Ng, Y. T. S., Chan, K. S., ... Leung, Y. M. A. (2009). *Elderly Commission's Study on Residential Care Services for the Elderly Final Report*. Hong Kong, China: Elderly Commission.

Connor-Smith, J. K., & Compas, B. E. (2002). Vulnerability to social stress: Coping as a mediator or moderator of sociotropy and symptoms of anxiety and depression. *Cognitive Therapy and Research, 26,* 39–55.

Crowther, M. R., Parker, M. W., Achenbaum, W. A., Larimore, W. L., & Koenig, H. G. (2002). Rowe and Kahn's model of successful aging re-visited: Positive spirituality—the forgotten factor. *The Gerontologist, 42*(5), 613–620.

Depp, C. A., & Jeste, D. V. (2006). Definitions and predictors of successful aging: A comprehensive review of larger quantitative studies. *American Journal of Geriatric Psychiatry, 14*(1), 6–20.

Fernandez-Ballesteros, R. (2011). Positive ageing: Objective, subjective, and combined outcomes. *Electronic Journal of Applied Psychology, 7*(1), 22–30.

Folkman, S., Lazarus, R. S., Pimley, S., & Novacek, J. (1987). Age differences in stress and coping processes. *Psychology and Aging, 2*(2), 171–184.

Freeman, L. C. (2004). *The development of social network analysis*. Vancouver: Empirical Press.

Fung, H. H., Stoeber, F. S., Yeung, D. Y. L., & Lang, F. R. (2008). Cultural specificity of socioemotional selectivity: Age differences in social network composition among Germans and Hong Kong Chinese. *The Journals of Gerontology, Series B: Psychological Sciences and Social Sciences, 63*(3), 156–164.

Gergen, M. M., & Gergen, K. J. (2001). Positive aging: New images for a new age. *Ageing International, 27*(1), 3–23.
Gui, S. X., & Huang, L. R. L. (2003). *A comparative study on social policy between Shanghai and Hong Kong*. Shanghai: East China Normal University Press. (in Chinese).
Hill, R. D. (2005). *Positive aging: A guide for mental health professionals and consumers*. New York: W. W. Norton & Co Inc.
Hill, R. A., & Dunbar, R. I. M. (2003). Social network size in humans. *Human Nature, 14*(1), 53–72.
Hong Kong Policy Research Institute. (2003). A sense of security, a sense of belonging and a feeling of worthiness for the elderly. *Policy Bulletin, 23*. (in Chinese).
Hong Kong Government. (1997). *Policy address by the chief executive*. Hong Kong: Government Printing Office.
House, J. S. (1981). *Work stress and social support*. MA: Addison-Wesley Longman.
Jonker, A. A. G. C., Comijs, H. C., Knipscheer, K. C. P. M., & Deeg, D. J. H. (2009). The role of coping resources on change in well-being during persistent health decline. *Journal of Aging and Health, 21*(8), 1063–1082.
Kahn, R. L., & Antonucci, T. C. (1980). Convoys over the life course: Attachment, roles and social support. In P. B. Baltes & O. Brim (Eds.), *Life-span development and behavior* (Vol. 4, pp. 253–286). New York: Academic Press.
Korchmaros, J. D., & Kenny, D. A. (2001). Emotional closeness as a mediator of the effect of genetic relatedness on altruism. *Psychological Science, 12*(3), 262–265.
Lai, W. L. (2009). From burden to depressive symptoms: The case of Chinese-Canadian family caregivers for the elderly. *Social Work Health Care, 48*(4), 432–449.
Lam, L. C., Lee, J. S., Chung, J. C., Lau, A., Woo, J., & Kwok, T. C. (2010). A randomized controlled trial to examine the effectiveness of case management model for community dwelling older persons with mild dementia in Hong Kong. *International Journal of Geriatric Psychiatry, 25*(4), 395–402.
Lan, P. C. (2002). Subcontracting filial piety elder care in ethnic Chinese immigrant families in California. *Journal of Family Issues, 23*(7), 812–835.
Lang, F. R., & Carstensen, L. L. (1994). Close emotional relationships in late life: Further support for proactive aging in the social domain. *Psychology and Aging, 9*(2), 315–324.
Lau, A., Chi, I., & McKenna, K. (1998). Self-perceived quality of life of Chinese elderly people in Hong Kong. *Occupational Therapy International, 5*(2), 118–139.
Lazarus, R. S., & DeLongis, A. (1983). Psychological stress and coping in aging. *American Psychologist, 38*(3), 245–254.
Lazarus, R. S., & Folkman, S. (1984). *Stress appraisal, and coping*. New York: Springer.
Legislative Council. (2011). *Proposal of the panel on manpower to send a delegation of two members to the Republic of Indonesia* (Ref: CB2/PL/MP). Hong Kong SAR.
Leung, K. K., Chen, C. Y., Lue, B. H., & Hsu, S. T. (2007). Social support and family functioning on psychological symptoms in elderly Chinese. *Archives of Gerontology and Geriatrics, 44*(2), 203–213.
Montross, L. P., Depp, C., Daly, J., Reichstadt, J., Golshan, S., Moore, D., …, & Jeste, D. V. (2006). Correlates of self-rated successful aging among community-dwelling older adults. *American Journal of Geriatric Psych, 14*(1), 43–51.
Moore, J. (2000). Placing home in context. *Journal of Environmental Psychology, 20*(3), 207–217.
Nagchoudhuri, M., McBride, A. M., Thirupathy, P., Morrow-Howell, N., & Tang, F. (2005). Maximizing elder volunteerism and service: Access, incentives, and facilitation. *The Journal of Volunteer Administration, 23*(1), 11–14.
Ng, S. H., Cheung, C. K. J., Chong, A. M. L., Woo, J., Kwan, A. Y. H., & Lai, S. F. Y. (2011). Aging well socially through engagement with life: Adapting Rowe and Kahn's model of successful aging to Chinese cultural context. *The International Journal of Aging and Human Development, 73*(4), 313–330.
Okun, M. A., & Stock, W. A. (1987). Correlates and components of subjective well-being among the elderly. *Journal of Applied Gerontology, 6*(1), 95–112.

Population Division of the United Nations. (2002). *Department of economic and social affairs, UN*. Retrieved online from http://www.un.org/esa/population/publications/WPA2009/WPA2009-report.pdf.
Rathus, J. H., & O'Leary, D. (1997). Spouse-specific dependency scale: Scale development. *Journal of Family Violence, 12*(2), 159–168.
Reichstadt, J., Depp, C. A., Palinkas, L. A., Folsom, D. P., & Jeste, D. V. (2007). Building blocks of successful aging: A focus group study of older adults' perceived contributors to successful aging. *American Journal of Geriatric Psychology, 15*(3), 194–201.
Rowe, J. W., & Kahn, R. L. (1997). Successful aging. *The Gerontologist, 37*(4), 433–440.
Ryff, C. D. (1982). Successful aging: A developmental approach. *The Gerontologist, 22*(2), 209–214.
Schnittker, J., & John, A. (2007). Enduring stigma: the long-term effects of incarceration on health. *Journal of Health and Social Behavior, 48*(2), 115–130.
Schwarzer, R., & Leppin, A. (1991). Social support and health: A theoretical and empirical overview. *Journal of Social and Personal Relationships, 8*(1), 99–127.
Social Welfare Department. (2014).*Support teams for the elderly*. Retrieved online from http://www.swd.gov.hk/en/index/site_pubsvc/page_elderly/sub_csselderly/id_supporttea/.
Tan, J., Ward, L., & Ziaian, T. (2011). Comparing definitions of successful ageing: The case of Anglo- and Chinese-Australians. *Electronic Journal of Applied Psychology, 7*(1), 15–21.
The World Bank. (2010). *World development indicators*. Retrieved December 2010 from http://data.worldbank.org/data-catalog/world-development-indicators?cid=GPD_WDI.
Vasquez, P. (2010). *Religious coping and social support as mediators and/or moderators and acculturative stress in a latino community sample*. Unpublished Master thesis, Marquette University.
Vaux, A. (1988). *Social support: Theory, research, and intervention*. New York: Praeger.
Walker, A. (2002). A strategy for active ageing. *International Social Security Review, 55*(1), 121–139.
Wenger, G. C. (1984). *The supportive network: Coping with old age*. London: Allen & Unwin.
Western, B. (2002). The impact of incarceration on wage mobility and inequality. *American Sociological Review, 67*(4), 526–546.
Wong, D. F. K. (2008). Differential impacts of stressful life events and social support on the mental health of mainland Chinese immigrant and local youth in Hong Kong: A resilience perspective. *British Journal of Social Work, 38*(2), 236–252.
Woo, J., Ng, S. H., Chong, A. M. L., Kwan, A. Y. H., Lai, S., & Sham, A. (2008). Contribution of lifestyle to positive ageing in Hong Kong. *Ageing International, 32*(4), 269–278.
World Population Ageing. (2009). *Population division, department of economic and social affairs, United Nations, New York*. Retrieved from http://www.un.org/esa/population/.
World Health Organization, Ageing and Life Course Unit. (2008). *WHO global report on falls prevention in older age*. Geneva: WHO.
Yeung, G. T. Y., & Fung, H. H. (2007). Social support and life satisfaction among Hong Kong Chinese older adults: Family first? *European Journal of Ageing, 4*(4), 219–227.

Chapter 2
Literature Review

2.1 Overview of Literature Review

This chapter synthesizes a wide spectrum of concepts, theories and previous academic research that guide the current research. It covers a review of concepts in social gerontology and psychology that concern the key concepts of the study, including positive aging, social support network, coping and DHs as support providers. Afterwards, a few concepts (e.g., successful aging, subjective well-being and quality of life) that may be related to the key concepts are also covered to give justifications of the choice of research topic in the specific study. Furthermore, a body of previous academic research in the relevant research area is presented. This review will enable the researcher to formulate conceptual frameworks for the study and to propose the hypotheses to be tested.

The chapter is composed of four parts. Part One is to introduce the research approaches of aging and specifically to elaborate the research focus—positive aging—in the present study. Part Two reviews the literature on social support network, including its structural and functional properties, as well as emotional closeness. In Part Three, coping is elaborated in terms of its definition, strategies and significance in late life. Part Four is to introduce the history and current situation of DHs in Hong Kong, and the role of DH as support provider in elder care is also discussed.

2.2 Research Approaches of Aging

In this section, positive aging as the dependent variable of the study is elucidated. In the meantime, the similar concept successful aging is also introduced. Additionally, two common approaches to explore well-being in late life, namely, subjective well-being and quality of life, are discussed.

S. Liu, *Social Support Networks, Coping and Positive Aging Among the Community-Dwelling Elderly in Hong Kong*,
https://doi.org/10.1007/978-981-13-3783-3_2

2.2.1 *Positive Aging and Successful Aging*

Positive aging and other relevant concepts including successful aging, active aging, and productive aging are currently regarded as the prevalent gerontology approaches to describe and explore older people's later life (Fernandez-Ballesteros, 2011; Hill, 2005; Reichstadt, J., Depp, C. A., Palinkas, L. A., Folsom, D. P., & Jeste, D. V., 2007; Tan, Ward, & Ziaian, 2011). Some social gerontologists argue that these terms are largely similar in their definitions, implications and conceptual framework, which all contain a couple of factors in depicting a kind of good aging process (Bowling, 1993; Bowling & Iliffe, 2006; Rowe & Kahn, 1997). Nevertheless, in the specific study, over reviewing numerous past literature on aging concepts, the researcher did not incline to classify these terms into the same category from a semantic perspective. As a matter of fact, distinct diversities between them could still be recognized and distinguished when further exploration and comparison are conducted. Thus, in this section, in addition to the elaboration towards my research focus—positive aging, successful aging will also be prioritized to be thrown light upon, since its similarity with positive aging is apparent and crucial, as well as its significant position in aging studies.

2.2.1.1 Positive Aging: Earlier Definitions and Descriptions

In general, positive aging usually represents such an umbrella idea or a guiding framework in order to describe the challenges and opportunities that can turn up in later life, and the role of individual choice in directing how one approaches growing old in this modern times (Hill, 2005). Another previous definition pointed out that positive aging may refer to the attainment of life satisfactions, coping behaviors and physical ability, which included the retention of social support networks that had been associated with well-being (Bowling, 1993). On the whole, the definition of positive aging is often related with an optimistic attitude towards the decline of physical and mental function as age increases. It has implications in many aspects of aging life, such as health issues, subjective attitudes towards life, and probable contribution to family and society (Chong, Ng, Woo, & Kwan, 2006; Fernandez-Ballesteros, 2011).

In view of the previous introduction of positive aging, the relevant literature was not that abundant since the theoretical and empirical studies on this specific theme were respectively rare. Table 2.1 provides a couple of descriptions and standpoints on positive aging from different perspectives aging.

In summary, the implication of positive aging in my study concerns the process of aging, not a reflection of a single state or result of aging. Positive aging is deemed as a series of dynamic attitudes and actions which are beneficial to construct a healthy and stable sunset life. In a word, under my original intention of this specific study, positive aging could be described as a kind of aging process requiring a person make choices, cope with stress, and continue contributing to family and society, even facing with the potential physical and psychological limitations that may come across in late life.

Table 2.1 Theoretical and empirical views on positive aging

Author	Research topic	Components and indicators
Bowling (1993)	Literature review on positive and successful aging	Attainment of life satisfactions, coping behaviors and physical ability
Hill (2005)	Characteristics of positive aging	Coping with age-related decline; making lifestyle choices to preserve well-being; cultivating flexibility across the life span; focusing on difficulties of growing old
Chong et al. (2006)	Views of positive aging	Physical health, cognitive functioning, positive attitudes about future, participation in society, and financial security, etc.
Cheung et al. (2002)	Conceptions of positive aging	Positive resources and positive contributions

Another similar concept related with positive aging is successful aging. For the term successful aging, it was once created by Rowe and Kahn (1998) to describe those older individuals who were able to live beyond the average life expectancy and experience quality aging into the seventh and eighth decades of life. The two concepts are often discussed together, because they all reflect a way or pattern how elderly people attempt to cope with the adverse effects during the process of becoming older. Their similarities mainly lie in the attitudes towards declined physical, functional and mental health brought by the added age. Both of them have a combination of physical and functional health, high cognitive functioning, financial security and active involvement with society (Chong et al., 2006).

Researchers have considered positive aging and successful aging could be regarded as equivalent concepts in gerontology. From the viewpoint of Bowling (1993), there is no distinct difference between positive and successful aging. However, from the superficial explanation of the two word "positive" and "successful", distinguishing diversities could be detected. "Successful" emphasizes the prosperous achievement, especially the attainment of something; while "positive" tends to be characterized by presence and have an affirmative nature (The Shorter Oxford English Dictionary, 1990). Therefore, in gerontology, successful aging attaches more emphasis on the successful achievements attained by the older adults when faced with a series of declinations. Furthermore, successful agers are probably defined as people who can maximize their life expectancy through controlling extrinsic factors (Rowe & Kahn, 1998). Nevertheless, the implications of positive aging, does not require that the maximization of life expectancy should be included as a main component. From the perspective of constructivism, it reveals that successful aging movement is wedded to an empiricist view of science, and positive aging is constructionist in its orientation (Gergen & Gergen, 2001). Then the differentiation between them is more distinct, successful aging probably addresses the outcome towards aging, including

Table 2.2 Brief comparison between positive aging and successful aging

	Positive aging	Successful aging
Orientation	Process-oriented	Outcome-oriented
Components (dimensions)	Good health, having a positive life attitude, active engagement with an activity or with society, feeling supported by their families and friends, being financially secure, and living in a place with emotional ties (Chong et al., 2006)	Low probability of disease and disease-related disability, high cognitive and physical functional capacity, and active engagement with life (Rowe & Kahn, 1997)

a healthy, long and satisfactory life (Lawton, 1983). However, from my perspective, as a rarely mentioned concept in empirical studies before, which turns up not as early as successful aging, positive aging attaches more importance to the aging process according to the elders' own construct in mind. Thus, this difference could not be neglected in defining the two concepts. In addition, the present study plans to explore how old people cope with the stressors brought by aging by virtue of their social support networks, especially the optimistic initiative, measures to face with the deficits. Thereby, compared with successful aging, the concept of positive aging was employed because of its most appropriateness and coincidence with the original research purpose. In Table 2.2, subtle differences could be recognized through a brief comparison between the empirical findings of both concepts. A more systematic elaboration on the origin and development on successful aging will be laid out in next section.

2.2.1.2 Successful Aging

Previous approaches

Successful aging issues have long been discussed for decades. In the earlier times, researchers proposed to define successful aging from several dimensions. A former statement was put forward from the perspective of life satisfaction in the Kansas City Studies of Adult Life (Ryff, 1982). Four dimensions of successful aging were presented, namely, the amount of activity which individuals engaged, the ability to disengage, satisfaction with life, and maturity or integration of personality. From the satisfaction orientation, successful aging has been explored in terms of happiness, subjective well-being, balance between aspirations and achievements, and other related psychological perceptions (Ryff, 1982). In light of this perspective, a vast of aging research was generated by this formulation, so that life satisfaction had long been the most essential research question in the area of successful aging.

In addition, other researchers tend to describe successful aging from the developmental and transactional perspective. This approach highly emphasizes the identification of conditions that lead to growth, including a series of strategies such as coping, optimization and so on. In Erikson's lifespan developmental theory, the final stage of ego development was *integrity* versus *despair* (Erikson, 1959). Thus, in light of this, emotional integration, accepting one's life cycle as something that had to be, making oneself adapted to the satisfactory experiences and disappointments were considered as predictors of successful aging. Afterwards it was further concluded as a growing process where older people sought for a higher and more differentiated maturity between the developing person and changing social context, rather than the pure satisfaction of life and non-developmental aging. Another social psychological approach on the topic of successful aging focused on positive coping (Fozard & Popkin, 1978). One of the objectives was to optimize older adults' abilities to adapt to various aging process from vision, memory to learning spheres, as well as to maximize individual choices related to health and leisure.

To conclude briefly, several approaches to successful aging from the life satisfaction, life-span developmental and social psychological perspectives are almost similar in the logic of defining. They all highlight the adaptive competence, development in the process of aging, rather than stress too much on the outcome such as life longevity, perfect health conditions.

Rowe and Kahn's approach

The formal and popularized term of successful aging was created by John Rowe and Robert Kahn, which could be dated back to an earlier comparison between usual and successful aging process (Rowe & Kahn, 1987). At that time, Rowe and Kahn were fully conscious of the distinction between usual and successful aging, which stimulated them to further explore the meaning of successful aging. In their statement, traditional aging research might focus too much on the morbidity and mortality issues, and then underestimated the extrinsic factors that could have a positive or neutral influence on aging process (Rowe & Kahn, 1987). On the basis of this point, they discussed several psychosocial factors such as social support and autonomy, together with their correlation with well-being. At last they concluded that it was the extrinsic factors that transformed usual aging to successful aging and that aging research should link the psychosocial and physiologic levels together. From my own perspective, Rowe and Kahn's first approach on the concept of successful aging was not comprehensively formed. They had pointed out a new way to carry out gerontological research—to pay more attention to psychosocial factors and the subjective well-being of older people, nevertheless, no concrete and direct definition of successful aging was presented.

Ten years later, they published an article which systematically delivered the definition and components of successful aging on the basis of the research outcome from *MacArthur Foundation Research Network on Successful Aging*. This time, they added a series of contents which had been often examined in usual aging—diseases and disabilities. The definition of successful aging included three main components, namely, low probability of disease and disease-related disability, high cognitive and physical functional capacity, and active engagement with life (Rowe & Kahn, 1997). The three parts of successful aging have their dimensions respectively. Low probability of diseases means not only the absence of illness, but the absence of potential risk of disease. As to high functional level, it attaches more emphasis on the potential and competence an old person could achieve in later activities. Lastly, active engagement with life most concern two factors—interpersonal relations and productive activity. It could be inferred that successful aging values the behaviors of older people which may be productive and create kinds of social value (Rowe & Kahn, 1997). Also, apart from intrinsic factors, extrinsic environment factors (e.g., elements of lifestyle) are addressed in the content of successful aging. It is apparent that this time Rowe and Kahn's successful aging attaches much importance on health issue. It holds that the maintenance of good functional and physical health is a distinct characteristic in response to normal aging process. In many empirical studies, the situation of "four Ds" (disease, disability, death and dementia) was regarded as an important criterion to measure successful aging (Strawbridge, Cohen, Shema, & Kaplan, 1996).

This statement of successful aging is more objective, since it addresses older adults' current situation of physical and functional health, as well as the participation of activities and social engagement, so that this kind of approach is most widely applied in gerontology studies. However, this definition of successful aging did not directly list extrinsic factors such as social support and control in their three dimensions, instead, continuing engagement with life was emphasized, which might be interpreted as a kind of integration of non-health issues. What's more, compared with their first publication, there is also a distinction that the psychological and intrapersonal factors are not highly stressed. So other researchers claimed that this model did not contain the spirituality which could produce positive affect to successful aging (Crowther, Parker, Achenbaum, Larimore, & Koenig, 2002). Otherwise, the implication of this model is also considered unrealistic because a disease-free older age is really an unapproachable fact (Bowling & Dieppe, 2005).

Following approaches

Posterior to the article on successful aging published by Rowe and Kahn (1997), other researchers also presented their ideas in regard to successful aging. Schemidt defined successful aging as minimal interruption of usual function, although minimal signs and symptoms of chronic disease might be present (Strawbridge, Wallhagen, & Cohen, 2002). So this approach has expanded the range of older people who could be called successful. In addition to this, another definition describes successful aging as doing the best with what one has, which, to some extent, also allows for light or chronic diseases (Baltes & Carstensen, 1996). Generally these new approaches still emphasize the importance of health and avoidance of diseases, but only reduce the level of high health-function demand. So it seems that these statements reveal few distinct differences from the construct of Rowe and Kahn.

Measure

Scholars have used various tools to measure successful aging in past empirical studies. Researchers inclined to construct measurement tools by virtue of the definition by Rowe and Kahn (1997). According to the features described by Rowe and Kahn (1998), researchers formulated responding scales or questionnaires to test each dimension of successful aging (Strawbridge et al., 1996). Several frequently-used variables and scales are presented in this section.

The first dimension of successful aging could be tested from these criteria below. For absence of disease, the absence of heart disease, stroke, bronchitis, diabetes, cancer, osteoporosis, emphysema, or asthma is included. Absence of disability included being able to perform all 7 activities of daily living (bathing, dressing, eating, using the toilet, moving from bed to chair, grooming, or walking across a room). Absence of risk factors included the absence of cigarette smoking, hypertension, and obesity, defined according to the National Heart, Lung, and Blood Institute's guidelines as a body mass index of 30 or greater based upon reported height and weight (Strawbridge et al., 2002). For the second dimension of successful aging, Townsend's physical functioning (e.g., activities of daily living) was widely adopted (Bowling & Iliffe, 2006). In terms of the third dimension—active engagement with life, researchers tended to include reporting monthly contact with three or more close friends or relatives. Being productive included reporting any of the following: (a) paid employment, (b) caring for a child or grandchild, (c) active volunteering, or (d) cleaning house (Strawbridge et al., 2002). In addition to these, General Health Questionnaire-12, and life satisfaction scale are also used to evaluate the psychological level. In a word, it could be concluded that the measurement of successful aging encompasses a large part of health-related index. And usually a combination of scales and questionnaires is necessary in accordance with the specific definition and interpretation of successful aging.

2.2.1.3 Revisitation of Positive Aging

In view of the previous two sections stating the definitions, origins and development of positive aging and successful aging, based on the comparison between both concepts, this section revisits positive aging through a recent two-factor model of successful aging and empirical studies in Chinese societies and finally conceptualizes positive aging in the present study.

A two-factor model of successful aging

In 2010, Pruchno, Wilson-Genderson, and Cartwright proposed a two-factor model of successful aging that included objective and subjective components. It is recorded in previous sections that measures of successful aging are diverse, including some that are single item and others that are multi-item and some that use objective measurements and others that rely on subjective assessments (Pruchno, Wilson-Genderson, & Cartwright, 2010). For the reason that successful aging has been endowed with objective criteria such as few chronic diseases, good physical and functioning health, and psychosocial factors, also known as subjective assessments, such as life satisfaction, attitude or appraisal of aging experience, Pruchno and associates constructed the two-factor model based on earlier conceptualizations of successful aging (Baltes & Baltes, 1990; Rowe & Kahn, 1987, 1998; Young, Frick, & Phelan, 2009).

In this model, the objective indicators include ample functional abilities, little or no pain and few chronic diseases; the subjective assessment includes attitude or appraisal of aging experience (e.g., the extent to which they rate their current life as positive, how well they are aging). Relevant empirical studies provided support to this construct of successful aging by a confirmatory factor analyses (CFA) of a survey conducted in the US. The two-factor model answers a critical question that whether older people can experience chronic diseases and functional declines and still feel that they are aging successfully. In other words, Pruchno and associates believe that they can. Therefore, this conceptualization of successful aging is much similar with the proposition of positive aging in the present study, that is, senior citizens can still consider themselves positive or successful even though they are experiencing physical or functional declines brought by aging process. Consequently, the implications of the two-factor model of successful aging build up the cornerstone of the conceptualization of positive aging in this study. In terms of the operationalization of the positive aging concept, this study borrowed one subjective indicator of the two-factor model, namely, life satisfaction, as a component to examine.

Additionally, it should be specifically pointed out that though the current research adopts the term *positive aging*, it embraces lots of meanings originated from *successful aging* in essence. Therefore, the following literature review also takes account of the empirical studies on successful aging.

Positive aging in Chinese societies

Numerous empirical studies have been conducted to examine the achievement of positive aging and successful aging in different parts of the world. Chinese societies in Asia are appropriate sites for research on positive aging, as life expectancy there are among the highest in the world and are still increasing. Cheung et al. (2002) conducted a research on the conceptions of positive aging among older people in Hong Kong. A total of 157 Hong Kong Chinese were divided into 12 focus groups, each of which contained 7 to 16 older adults. Then questions related to family, environment, career, current life, finance, and health, etc. were distributed to collect perceptions on positive aging. Consequently, their perceptions about positive resources mainly included autonomy, self-reliance, knowledge and ability of self-work, and so on. The positive contributions reported in this study covered socializing, stress management, experiencing support on self-level and working, generativity, and volunteerism on society-level. Later on, Chong et al. (2006) conducted a qualitative study of older adults in Hong Kong wherein respondents were asked a series of questions concerning their perception toward positive aging. Results indicated that the components of positive aging were identified as good physical health and cognitive functioning, a positive attitude about the future, support from family or friends, participation in society, financial security, and living in a place with emotional ties.

Although a number of researchers have adopted Rowe and Kahn's model in aging studies among older Chinese (Li et al., 2006), there is a lack of culturally sensitive measurement tool designed from the indigenous views of successful aging from Chinese older adults. Scholars in Hong Kong have recently attempted to explore successful aging with the use of both qualitative and quantitative methods. Inspired by the implications of the study conducted by Chong and associates and the theoretical background of successful aging proposed by Rowe and Kahn (1997), Ng et al. (2011) reinterpreted successful aging with reference to the unique characteristics of

Chinese folklore and constructed a new framework. Among the classic dimensions of successful aging outcomes, engagement with life was elucidated to encompass two elements, namely, caring engagement (CE) and productive engagement (PE). Derived from traditional Chinese culture, CE embodies a popular Chinese cultural notion that is grounded in a Confucian ethic of caring for young ones and for elders in one's own family as well as in other families (*you wu you yi ji ren zhi you; lao wu lao yi ji ren zhi lao*) (Ng et al., 2011). Hence, engagement with others by showing care and love to people related or known to oneself (e.g. family members, friends, relatives, colleagues, etc.) is regarded as a form of social engagement in Chinese society (Pang-White, 2009). Additionally, PE reflects another Chinese value that elders have roles to play (*lao you suo wei*), which suggests that the elderly could generate productive contributions to society. For instance, undertaking paid or non-paid work, social and political participation and volunteering are forms of PE. These two discursive themes underpinning engagement with life have been validated in empirical studies (Ng et al., 2011). Thereby, the term engagement with life under the classic framework of successful aging has been endowed with new essence specifically relevant to Chinese older adults. Hence, the researcher will adopt the four components of Ng et al.'s construct as the objective criteria of positive aging in this current research.

2.2.2 *Subjective Well-Being and Quality of Life*

2.2.2.1 Subjective Well-Being: Definitions and Components

Subjective well-being (SWB) is a popular concept which is extensively investigated in the field of social gerontology. In the viewpoint of many practitioners, measures could be done to improve the older persons' quality of life via enhancing their subjective perception of well-being (Okun & Stock, 1987). In the earlier time, Wilson (1967) once gave a description of well-being by concluding a happy person as a "young, healthy, well-educated, well-paid, extroverted, optimistic, worry-free, religious, married person with high self-esteem, high job morale, modest aspirations, of either sex and of a wide range of intelligence" (p. 294). Since then, researchers have investigated a lot in SWB and the results from theoretical and empirical studies are really influential.

Although the definition of SWB varies in the academia, one created by Okun is widely recognized. It goes to that SWB refers to the affective reactions of individuals to their life experiences along a positive-negative continuum (Okun, 1995). Referring to this definition, key indicators of SWB have also been elaborated as happiness, life satisfaction, and morale (George, 1979). Another brief introduction about SWB can be simply defined as the individual's current evaluation of his or her happiness. This kind of evaluation is often expressed in affective terms, for example, when asked about SWB, participants will often say, "I feel good" (Schwarz & Strack, 1999). Because the issue of well-being is mostly related to one's subjective perception and opinions, a majority of definitions are created to reflect the attitudes towards life of the respondents. Also, with the application of social statistics in social science field, other scholars attempt to define SWB through factor analysis concerning the structure of SWB, which promotes the development of exploring the indicators and components.

The component of subjective well-being was defined as a theoretical construct and represented a necessary part of the explanatory network for SWB, which had either structural or process functions (Okun & Stock, 1987). Liang (1985) had tested the structure of SWB via factor analysis and pointed out congruence, happiness, positive effect, and negative effect were the four first-order factors. Later on, Stock, Okun, and Benin (1986) proposed their model of components of SWB, which consisted of positive affect, negative affect, and cognition components. In another theoretical study towards the components of SWB (Deiner, Suh, Lucas, & Smith, 1999), SWB was defined as a broad category of phenomena which includes people's emotional responses, domain satisfaction, and global judgments of life satisfaction.

For the definition of SWB, the viewpoint of Deiner et al. (1999) was relatively sound and detailed. It contained the facets of respondents' attitudes of well-being in a specific level. Regarding to the components of SWB, the classification by Stock et al. (1986) is more comprehensive and sententious, which is relatively easier to be measured.

2.2.2.2 SWB: Theoretical Findings and Measures

Wilson (1967) established a theoretical model to explain the individual differences in terms of SWB. Diener (1984) explained that theory and elaborated from the perspective of a bottom-up approach. It stated that there were basic and universal human needs and that if one's circumstances allowed a person to fulfill these needs, he or she would be happy, on the contrary, the persistence of unfulfilled needs caused unhappiness. In light of this theory, other researchers have conducted many studies and found that demographic factors such as sex, age, race, education, and marital status were not the main leading factors of the variance in SWB.

Another group claimed that personality was one of the strongest and most consistent predictors of subjective well-being. Relevant studies according to this theory mainly focused on exploring the genetic association with SWB. In a reanalysis of Tellegen et al.'s twin study, Lykken and Tellegen (1996) calculated that although 40–55% of the variation in current SWB can be explained by genes, 80% of long-term SWB is heritable. Based on the later heritability estimate, it could be said that it is as hard to change one's happiness as it is to change one's height (as cited in Diener et al., 1999). Another finding indicated a hypothesis that extraverts were happier than introverts because of greater personality-environment fit (Diener et al., 1999). This finding led a series of attention on improving the social situations and individual involvements in society; consequently, the knowledge of subjective well-being was enhanced.

Additionally, Michalos (1985) proclaimed a theory of SWB that individuals tended to make a comparison between themselves and multiple standards which

Table 2.3 Definition and components of subjective well-being

Author	Definition	Components
Okun and Stock (1987)	The affective reactions of individuals to their life experiences along a positive-negative continuum	Positive affect, negative affect and cognition
Schwartz and Strack (1999)	Individual's current evaluation of his or her happiness	NA
Deiner et al. (1999)	A broad category of phenomena that includes people's emotional responses, domain satisfactions, and global judgments of life satisfaction	Emotional responses, domain satisfactions, and global judgments of life satisfaction
Liang (1985)	Nil	Congruence, happiness, positive effect, and negative effect

included past conditions, experiences, other people and other needs or goals. Judgments about satisfaction were drawn from these comparisons. The feeling of discrepancy became the leading factor of decreasing SWB. The discrepancy theory inclines to analyze the situation of SWB from the angle of people's attitudes of past events and the process of comparing with others. In addition to that, relevant studies on goals and needs affecting SWB are also derived from this theory.

As is recognized, subjective well-being is a complex entity which contains multiple facets. However, it does not mean no measurement could be carried out on SWB. And it seems to be widely believed that SWB can be measured by physiological characteristics, behavioral reactions, and memories. In the study of Okun and Stock (1987), they recommended the Affect Balance Scale (Bradburn, 1969) as a measure of happiness; three items (12, 13, and 19) from the Life Satisfaction Index A (Neugarten, Havighurst, & Tobin, 1961) as a measure of life satisfaction; and the Hope Scale with slightly revised instructions as a measure of morale (as cited in Okun & Stock, 1987). Then the three measurement tools could be combined together as a scale of SWB. However, this kind of measure seems a bit confusing when adopting a combined scale from other three different scales, and it is not feasible in operational process. Additionally, there are other kinds of scales, for example, Ecological Momentary Assessment, and Day Reconstruction Method, etc. A brief chart (Table 2.3) on relevant knowledge about subjective well-being is listed below.

2.2.2.3 Quality of Life

Quality of life (QoL) is a concept frequently mentioned and applied in many scientific fields, including disciplines of gerontology, psychology, sociology, etc. It is a wide-range term because it usually involves multiple domains (health, psychological, social and environmental), containing both objective and subjective components (Fernández-Ballesteros, 2010). It is also a dynamic concept, reflecting values as they change with life experiences and the process of aging (Bowling & Gabriel, 2004). QoL has been utilized in individual level to assess individuals' health, subjective well-being, life satisfaction, and social well-being. In addition, it could also reflect the general situation of a region or society in macro level, for instance, serving as a crucial social indicator in large-scale surveys.

As a multidimensional concept, the definition of QoL varies. Considering its wide use in measuring the living conditions of citizens in some region or area, quality of life has been defined by WHO (1993).

> ...an individual's perception of their position in life in the context of the culture and value systems in which they live and in relation to their goals, expectations, and standards and concerns. It is a broad-ranging concept affected in a complex way by the persons' physical health, psychological state, level of independence, social relationships, and their relationships to salient features of their environment.

From this definition, it indicates that the situation of QoL is much related to the social context and culture. The environment that individuals live in may have a direct influence of the outcome of the perception of QoL. Additionally, from the individual level, people's attitudes towards QoL might be clarified in the physical, psychological and social domains.

Other researchers also attempt to give a definition of quality of life. Although it is widely recognized that it is difficult to make a comprehensive definition, there are various types of definition from different perspectives. In light of the authors' statement in scientific domain, QoL has been defined as equivalent to well-being in the social domain (Campbell, 1981), to health status in the bio-medical field which uses the Health-related QoL (Naughton & Wiklund, 1993), and to life satisfaction from a psychological viewpoint (Palys & Little, 1983). However, since QoL is a multidimensional construction, solely exploring it from only one domain seems not convictive. For example, Neugarten et al. (1961) pointed out that quality of life can be wholly or largely determined by the use of life satisfaction situation, which remained further exploration. Thereby, many researchers considered to define QoL in terms of its components. Brown et al. has distinguished several types of QoL components, which are objective social indicators, satisfaction of human need, subjective social indicators, personal resources (e.g., social capital), ecological and neighborhood resources, health and functioning and psychological concepts (as cited in Fernández-Ballesteros, 2010). Other researchers also conducted similar studies in the past twenty years to explore the components or elements of QoL and consequently got a close outcome. Lau, Chi, and McKenna (1998) carried out a qualitative study in Chinese elderly people in Hong Kong to explore their perceived QoL; it concluded the elements of QoL, meanwhile classified all the elements into four domains—physical and functional well-being, psychological well-being, social well-being, and economic well-being.

To obtain a systematic understanding of QoL, Bowling and Gabriel (2004) established an integrated regression model of QoL in older age from a British survey. This study collected all the perceived definitions of QoL from the respondents and compared the results of an analytic regression model of theoretically derived indicators. And this model took different domains, levels and components of QoL into consideration and analysis, then generated a model, which could act as a reference in both descriptive and evaluative research. In this model, good relationships are the most commonly mentioned constituent of QoL; other important factors are social roles and activities, health, psychological outlook and well-being, home and neighborhood, finances, and independence. This study is significant and persuasive, because it provided empirical data on which theoretical models and elements of QoL can be examined with their high content validity. Another highlight of Bowling's model of QoL is that it focuses on more than the traditional social science model of life satisfaction or health, since in the earlier time, many researchers assumed that QoL resembled life satisfaction and well-being. So this study explicitly put forward a scientific model via statistics analysis, taking QoL as an independent and featured concept.

Compared with other relevant concepts-subjective well-being, life satisfaction, happiness, etc., QoL embraces its unique feature. That is to say, though there are many similarities in conceptualization, components and even measurement between QoL and well-being issues, QoL is not equivalent to those ones. As a kind of social indicator or outcome representing societal or individual living situations, it could not be totally transformed into the subjective appraisal of those conditions (Sampson, 1981). For its especial broad range of application in social studies, QoL is obliged to act as a systematic measurement concept from both objective and subjective approaches. That's the differentiation between QoL and other related concepts.

2.2.2.4 Summary

Quality of life, life satisfaction and subjective well-being are frequently studied in empirical studies. Frankly speaking, QoL, as a relatively broader concept, contains more information than the others. As a multifaceted concept, QoL embraces environmental factors and effects, as well as both the perceived and objective conditions of the respondents. For its wide boundary and complicated information contained, to apply quality of life in this study could not highlight the aim of concentrating on how to cope with aging. And since the present study will be carried out in individual level, the outcome of quality of life seems a bit huge in analysis, which could not indicate a direct, proper and precise situation corresponding to my research goal. In the same way, other concepts including life satisfaction and subjective well-being, though applied in gerontology many times before, are not created to study the elderly particularly. Meanwhile the results of life satisfaction and subjective well-being are almost depended on the respondents' perception to a large extent, so the assessment of aging process reveals a bit weak and lacking of persuasion. Lastly, in the past studies related to the elderly, empirical research on positive aging among Chinese older people was not frequently seen. Therefore, in summary, the present study would like to select *positive ag*ing as the research focus.

2.3 Social Support Network

In this part, a detailed description of the important concept—social support network in the present study is to be outlined. The original construct of this term and a few concepts with similar meanings are also presented.

2.3.1 Definition of Social Support Network

In view of the previous literature, the concept *social support network* has long been defined in a consistent way by scholars. Generally the term social support network is often closely interrelated with social support and social network. House, Umberson, and Landis (1988) offered the following definitions of the two terms. Social network referred to the structural properties that characterized a set of relationships. Social support referred to the emotionally or instrumentally-sustaining quality of social relationships. And based on this distinction, social support network is experientially disserted as the combination of the portions that shape a network for the sake of providing support to persons. Wenger (1984) claimed that individual's support networks derived from the larger social network and consisted of those who were available for the person to provide regular emotional support, instrumental help and advice.

Reviewing the previous studies on this aspect, the concrete and specific definition of social support network is rarely mentioned. Nevertheless, the internal logic of conceptualization is far from diverse. And scholars tend to regard social support network as a group of relationships including kinship, friendship, neighborhood etc., which could offer support to the person in various aspects. For the defining logic, there is no obvious dispute, although some different explanations towards the concept of social support may still exist. In a classical statement of the term social support by House (1981), it was once defined as the below types: emotional, instrumental, informational, and appraisal support. Lee, Ruan and Lai (2005) claimed that sources of support to the focal individual may include a variety of people or institutions, such as spouse, parents, children, siblings, friends, neighbors, colleagues, as well as the various kinds of professionals and work organizations. Apart from the two, other scholars also attempted to draw a definition on social support, which were broadly selected in Table 2.4.

Table 2.4 Definitions and components of social support

Author	Definition	Components
Lin, Simeone, Ensel, and Kuo (1979)	Support accessible to an individual through social ties to other individuals, groups, and the larger community	NA
Cobb (1976, 1979)	Information belonging to one or more of the following three classes: information leading the subject to believe that he is cared for and loved; that he is esteemed and valued; that he belongs to a network of communication and mutual obligation	Emotional support, esteem support, and network support and Instrumental support or counseling, active support or mothering, and material support or goods and services
Kahn and Antonucci (1980)	Interpersonal transactions that include one or more of the following key elements: affect, affirmation, and aid	Affect, affirmation, and aid
Caplan and Killilea (s)	Attachments among individuals or between individuals and groups that serve to improve adaptive competence in dealing with short-term crises and life transitions as well as long-term challenges, stressors, and privations	Professions and formal community institutions as well as natural systems
House (1981)	Emotionally or instrumentally-sustaining quality of social relationships	Emotional support, instrumental support, informational support, and appraisal support

See House (1981). Work stress and social support

In a synthesis, all the above descriptions about social support could be concluded as a network of family, friends, neighbors, and community members that is available in times of need to serve psychological, physical, and financial help. Consequently, the definition of social support network in the present study may be generated as a number of social support providers which make up of an integrated network.

Table 2.5 sums up several theoretical standpoints on social support network from the perspectives of four classical sociological theories.

For the correlation with positive aging or successful aging, previous studies have demonstrated that social support could play a positive role in the promotion of successful aging (Rowe & Kahn, 1987). Specifically, the condition of social support or social networks may affect the health status of old people (Blazer, 1982; Cohen &

Table 2.5 Theoretical views on social support network

Author	Theory	Viewpoint
Homans (1961)	Social exchange theory	The more rewards provided, the more supportive the relationship would be Sharing interactions and exchanges within a relationship are supportive
Festinger (1954)	Social comparison theory	People attain cognitive support from interpersonal relationships by comparing their own performances and opinions with others
Wills (1985)	Personal control theories	Social networks may enhance feelings of personal control because of their availability for providing needed aid and resources in times of crisis. It predicts that it is the perceived reliability of networks rather than any current exchanges that provides a supportive function

See Wills (1985). Supportive functions of interpersonal relationship

Syme, 1985), Additionally, it would also have a buffering effect to stress, which to some extent promotes the self-reported well-being and then further facilitate successful aging (Cohen & McKay, 1984; Rowe & Kahn, 1987; Russell & Cutrona, 1991). Thus, to explore the relationship between social support network and positive aging is of great significance to the elderly.

2.3.2 *Concepts in Relation to Social Support Network*

2.3.2.1 Social Network

In the academic field of sociology, the terminology social network has been constantly utilized in studies. It was first generated in the sociology domain, which is further developed and improved in other fields (e.g., information sciences, internet techniques), reflecting a considerable growth of interest in the academia. Accompanied with the social network theory, network analysis has been booming in the past several decades. For my specific research, the social network mentioned here points to the formal phrase applied in sociological research. The origin occurrence of social network may be dated back to 1954, Barnes begun to use the term systematically to denote patterns of ties, encompassing concepts traditionally used by the public and bounded groups (e.g., tribes, families) as well as social categories (e.g., gender, ethnicity) which were used by social scientists (Freeman, 2004).

In the following years, quantities of researchers have explored the social network concept, components, measurement and implications. The concrete definition of social network varies in the academic field from different dimensions. It could be generally defined as a social structure made up of individuals (or organizations) which are connected by one or more specific types of interdependency, such as friendship, kinship, common interest, financial exchange, sexual relationships, or relationships of beliefs, etc. In regard to the research focuses on this term, scholars inclined to examine the quantity and quality of social network in the past studies. Quantity properties of social network include the size, density, frequency of contacts and centrality, etc. (Hill & Dunbar, 2003; Otte & Rousseau, 2002; Scott, 2007); while the aspects of quality are usually measured by specific items such as getting emotional support, intimacy and satisfaction (Pinquart & Sorensen, 2000).

Many authors may confuse the concept social network since it might share a few similarities with the research content in the present study—social support network (Cohen & McKay, 1984). In terms of social network, this concept stresses the frequency of interaction and the importance of relationship with the target person, usually neglects the exchange of resources and reciprocal facts (Shinn, Lehmann, & Wong, 1984; Wellman, 1981). However, the essence of social support network highlights the exchange of resources intended by the provider or perceived by the recipient which are considered beneficial to the recipient (Shumaker & Brownell, 1983). Another statement about social support was described as the feedback provided through contacts with similar and valued peers. The evident distinction between the two concepts lies in the intrinsic affecting directions to the target person. Social network may derive from the social interactions and social contacts with others, without any presupposed goal and explicit communication needs. Thus, the relationships in social network may not always be supportive—they could strain individual's resources or make him (her) excessively independent, which might produce some hurt on his or her feelings (Ingersoll-Dayton, Morgan, & Antonucci, 1997).

Another interpretation towards social network is that it may embed both positive and negative effects; nevertheless, in the process of support provision, the expected effects are almost positive-oriented. Whereas, it is still argued that the social support models have negative outcomes, if the perception of support providers is taken into consideration, or reviewing from a cultural perspective. Providers may obtain intrinsic satisfaction and social approval for providing support, but they expend time, resources and energies in the process, leading to some latent negative feedbacks in the relationships (as cited in Shinn et al., 1984). But in the present study, the party of provider would not be involved in the research scope, and it is generally approved that the received support in one's social support network bears positive effects.

To conclude briefly, in this current study, the social support network and social network terminology could not be overlapped because of the internal diversities. It is apparent that the range of social network is conceptually broader than social support network, which contains more kinds of interrelationships. However, as a type of network, it is approachable to adopt the analytical methods to examine the social support network (e.g., types and size).

2.3.2.2 Social Capital

In the lately times, the concept of social capital has witnessed a significant development among the sociologists, political and economic scientists in the social science domains. As an integrated and comprehensive concept with multifaceted implications, studies focused on social capital have been widely carried out to explore the areas including families, youth behavior problems, education, community life and governance, etc. (Adler & Kwon, 2002). The social capital concept may represent certain differences in implications and the theoretical framework, compared with the concepts of social network, social relationships and social support network. However, it inherits the analytical methods from network theories, which calls for a rough description and differentiation in this study.

Social capital is generally understood as a goodwill which is embedded in the organization of social relationships and might be potentially mobilized to facilitate various targeted actions (Adler & Kwon, 2002). The specific definitions vary from different angles and dimensions. The origin systematic definition of social capital may derive from Pierre Bourdieu, which was "the aggregate of the actual or potential resources which are linked to possession of a durable network of more or less institutionalized relationships of mutual acquaintance or recognition" (Bourdieu 1985, p. 248; as cited in Portes, 1998). Afterwards, scholars preferred to make a definition fundamentally from the dimension of the sources, which were divided into three directions in general—the relations from external actors, the structure of relations within an internal collectivity, or both types of resources (Robison, Schmid, & Siles, 2002). Then another dominating definition of social capital was described as "Social capital is defined by its function. It is not a single entity, but a variety of different entities having two characteristics in common: they all consist of some aspects of social structure, and they facilitate certain actions of individuals who are within the structure" (Coleman, 1994, p. 302). Although social scientists might have come up many significant definitions with some nuances, their starting points in defining were similar, highlighting the owned existing relationships from social structure, as well as actors' purposes and actions to gain the achievements. Regarding the sources of social capital from my perspective, this concept could be broadly stated as an aggregation from the interrelations from actor's social structure which may be deemed to have already or will produce a crucial influence in future success and purposed achievements.

In the analytic framework of social capital, social scientists usually have more discussions on the forms, benefits and risks of the concept. It was suggested that social capital might be implanted in three dimensions (Chong & Ng, 2010). Bonding social capital, which was claimed as "by chance or necessity, inward looking and tend to reinforce exclusive"; while as for bridging social capital, was considered "outward looking and encompass people across diverse social cleavages" (Putnam, 2001, p. 22). Additionally, the third form of social capital was categorized as linking social capital, referring to the relations between individuals and groups in different social strata (Ting, 2010). And in detailed analysis, all the three types of social capital may have diverse influences to individual or community's purposed action.

In terms of the consequent benefits in the macro and micro levels, social capital could be demonstrated to have close association with a substantial number of information production, trust and social control, as well as good governance and solidarity (Adler & Kwon, 2002; Ting, 2010). Paradoxically, social capital could also generate a series of negative characteristics which might be regarded as risks. They are embedded in the below aspects—the undeserved and less efficient investment in certain situations (Adler & Kwon, 2002), high levels of intercommunity distrust and intolerance (Ting, 2010), etc.

Reviewing the two concepts—social capital and social support network, the focal similarity they show in common is that they both have some multidimensional implications in the internal construct. However, the diversities are salient in the components. With regard to social capital, the actors are universally endowed with explicit goals or purposes in implementing actions, which constitutes the vital reason of accumulating and developing social relationships in networks. In addition, the implication of social support network could be elaborated as various kinds of support provision deriving from the actual needs and demands. Thus, seemingly social capital has a profounder meaning than social support network and tends to be utilized in analyzing individual or social development. But for this present study, one of the main research variables has been determined as social support network, whose intrinsic implications and dimensions are more approachable to the research objectives, which could better reflect the pragmatic supportive needs of elderly citizens.

2.3.3 *Dimensions of Social Support Network*

In terms of the traditional analysis of social support, social scientists tend to clarify its structural parts and functional parts. In this section, a detailed elaboration from the two dimensions will be presented, in which my proposed research contents would also be generated. Additionally, emotional closeness, as a special property of social network, will be introduced.

2.3.3.1 Structural Properties

Types of social support network

Social support usually refers to the functions performed for the individual by significant others, such as family members, friends, neighbors and coworkers (Thoits, 1995). These support providers are types of relationships in a network, which could be interpreted as an indicator of the structure of social support. Network types have been identified as a key aspect of social support for its systematic variation among support elements across individuals (Cheng, Lee, Chan, Leung, & Lee, 2009). In previous studies, the type of support network of the older adults is largely determined by biology factors including siblings, children; migration patterns, such as the availability of family, friends; and the individual's own temperament (Kirby, Denihan, Bruce, Coakley, & Lawlor, 2000). Theoretically, the types of social support could be various and they do not have a strict boundary; while in fact, as to the older adults living in a fixed place, the common support providers of the old people are almost stable since they do not have to experience changes from family and career.

As a kind of diverse network, to define the components from the angle of structure, the explicit parts of support network people affiliate with could be listed as family members, friends, neighbors, professionals, and social organizations or institutions, etc. Lee et al. (2005) have clarified the types of support sources into three categories: close kin (e.g., spouse, children, parents and siblings), extended kin (e.g., uncles and aunts), non-kin primary groups (e.g., friends, neighbors and coworkers) and secondary groups (e.g., professionals, voluntary agencies and government departments). For the elderly, family members are the most familiar support providers to them, which consist of spouse, siblings, children and other relatives, etc. Additionally, an older person usually has some social interactions with others, of which the neighbors and friends are the most frequently communicated with. At the moment, more professionals appear in the living environment of elderly people. The involvement of domestic helper, as a common characteristic in Hong Kong households, is playing an important role in taking care of the old parents at home. Apart from that, hospital or clinic doctors, social workers, nursing workers tend to be indispensable to some older adults because of their support to them. What's more, nowadays many community centers, neighborhood centers and other social organizations have been established for the older people, which are considered to be a source of social support.

Family support has long been a research focus because of its unshakable role in relation to individuals' well-being in past literature (Chen & Silverstein, 2000; Cheng et al., 2009). Additionally, support from peer group members including friendship with center members, old classmates, colleagues, and church members is increasingly valued by community-dwelling elderly. Friendship in later life was demonstrated a prevalent factor that can contribute to psychological well-being and to expel loneliness in recent studies (Holmén & Furukawa, 2002; Victor, Scambler, Bowling, & Bond, 2005). Therefore, the present study will focus on family members and peer groups in social network analysis and elucidate their association with positive aging.

Size of social support network

Previous studies have explored the size of social relationships or social networks, which could serve as a predictor of the general range of a network (Freeman, 2004; Hill & Dunbar, 2003; Pinquart & Sorensen, 2000; Scott, 2007). Different from social network, whose size is measured in accordance with the frequency of contact and interactions (Hill & Dunbar, 2003), support network attaches more emphasis on the process of support exchange and those behaviors in which supportive interactions occur.

In my definition, the size of support network is the number of support givers who could provide different functions of social support to the elderly in need. In addition to the total number of support providers, the number of support providers in each type of support resource (including family members, relatives, friends, neighbors and professionals) will also be calculated respectively.

Thus, the size of social support network does not only mean the number of persons you have relationships with, it contains all the involved who can provide support in need and whose role could not be neglected derived from the perception of respondents. Social support network, as part of social network, its size could also be applied to indicate to what degree the elderly are involved in social context and the number of relatives, friends, and professionals, with whom contact is maintained (Van Tilburg, 1995). In addition, it was generally proposed that larger network size could be more effective to provide supportive behavior as well as information, accessibility, and perspectives, etc. (Vaux, 1988). While in another study, it was claimed that the larger the kin network (i.e., more adult children), the rate of social support got smaller (Son & Lin, 2006). Thereby, whether the statement is suitable in Chinese society especially in Hong Kong need further investigation.

2.3.3.2 Functional Properties

For a long time, the classification of social support functions is widely discussed. Scholars have carried out many studies to examine the different functional properties of social support (Barrera & Ainlay, 1983; House et al. 1988). From a review of prominent discussions of support functions, it was proposed that social support included five categories: material aid, behavioral assistance, intimate interaction, feedback, and positive social interaction (Barrera & Ainlay, 1983; House et al., 1988). However, considering the current situation in Chinese society, this classification could not be deemed as a comprehensive one. In light of the five categories, some overlapping meanings seem to exist between them. Intimate interaction and positive social interaction may share some similarities in practical analysis, leading to ambiguous judgments. Later on, it is generally accepted that social support contains emotional support, instrumental support, informational support, and appraisal support (House, 1981). Moreover, spiritual support is also mentioned in previous literature. Spiritual support provided by clergy, chaplains and other spiritual leaders may play an important role in spiritual care. In addition, family and friends can offer important spiritual support too. Spirituality refers to an inner, subjective region of life that revolves around individual experiences of being, transcending the personal self, and connecting with the sacred (Atchley, 2006). Receiving thoughtful spiritual support as people struggle with spiritual issues can make a significant difference in someone's life (Park, 2005). However, the present study did not aim to take spiritual support as the research focus because of the nature of the survey design. The limited time of completing a questionnaire might not allow collecting detailed data related to spiritual support. Nevertheless, in the modern time, this classification still reveals its drawbacks and inadequacy, especially for the specific research goal of studying elderly people. By virtue of the above two classifications, the study divides the support functions into material support, emotional support, companionship support, instrumental support and informational support.

Material support refers to provision of tangible sources in the form of money, materials which could be regarded as the economic basis of living. It should be noticed that material support here only covers the substantial and visible support, not include services and labor. Nowadays, with the rapid development of urbanization and industrialization, Hong Kong is undergoing lots of changes in life styles. Then older adults may have more choices when they are in need of material or financial support. In the previous statement about support functions, material support is not listed directly. But in reality, a majority of the older adults do not have any salary because they have retired, and the pension policy is not so extensive to cover every old person, so the financial resources are quite important in different modes of elderly care. In the meantime, material support is also a vital factor affecting old people's health and living conditions, which should be paid much attention to.

Emotional support, involves providing love, empathy, caring, listening and trust, as well as those providing support which could release the negative mood (House, 1981). It has been proved that the emotional support has a salient influence on the physical and mental health of the elderly (Cheng et al., 2009; Cohen & Syme, 1985; Schwarzer & Leppin, 1991). And in reality, emotional support could really help a person to improve self-esteem and expel unhealthy mood (Vandervoort, 1999). For the older people, when they are confronted with tough problems and worries, they may turn to different kinds of persons for psychological comforts, which could be understood as a kind of emotional support. And this support usually emerges together with desires, troubles, enquiries and some similar actual needs in specific situations.

Companionship support describes a kind of inner desire to be accompanied by others, from which to acquire the belongingness. For the elderly people, especially those who are living separately from their children or widowed, they tend to have a strong need to find a person to talk with to expel loneliness. This kind of support has reflected an increasing need with the rapid speed of urbanization and industrialization. In big cities such as Hong Kong, most of the adult children need to struggle for work and they usually have little time to look after their parents and only leave them at home. As a result, many older adults are reported to have strong loneliness, leading them to find companionship from others to spare their leisure time and further facilitate positive affective moods (Wan, Jaccard, & Ramey, 1996). It has been demonstrated that companionship support is very essential in predicting subjective well-being of Chinese older adult (Cheng et al., 2009). Additionally, in Hong Kong, domestic helper is a large group who provide services to a household, which are sometimes accepted as a family member. And this unique situation appears in Hong Kong suggests a new kind of companionship support for the elderly, since amounts of older people have looked on the domestic helpers as an essential role. Unlike the emotional support mentioned above, the elderly usually hope to seek for companionship without specific desires, nor in special occasions. They just treat this companionship as a constant part of their daily life, instead of looking for someone to give them actual immediate help and assistance. So the companionship support reflects its increasing importance in modern society. Considering its vital role in old people's daily life, therefore it is significant to take companionship support into the study.

Instrumental support could be defined as kinds of services, behaviors and other non-material forms of support to directly help the person in need. Instrumental support once has been served via the provision of material, financial assistance, suggestions and clarifying issues (Vaux, 1988). Different from what House (1981) and Vaux (1988) claimed, in this study, this kind of support does not include financial and material elements. It focuses on those behaviors such as taking care of the old, curing their diseases and so on. Theoretically, all the support providers could give help to old people when they are in trouble, which are generally considered as provision of instrumental support. Generally, in addition to the family members and acquaintances, social workers, doctors and other professionals could also provide instrumental support to the elderly who are in need. For those elderly people living in residential institutions, they may receive instrumental support from their caregivers. Therefore, instrumental support should be an important source of problem-focused help to the elderly.

Informational support means providing useful information which could be applied to solve problems. In contrast to instrumental support, information here is not of itself help, but helps people to cope with problems themselves (House, 1981). Informational support always acts as a medium to provide assistance and guidance to persons, such as necessary knowledge and skills to resolve various stressful situations (Cresci, 2001). Additionally, advice, cognitive guidance, feedback and direction, etc. are also labeled as part of informational support because of their function of assisting to solve problems (Cresci, 2001). At present, older adults are often encouraged to live independently with empowerment; in this case, informational support could reveal its benefits in helping older people. For those elderly who take amateur courses, or listen to health care lectures regularly, what they are receiving could be classified into information support.

To conclude, it is not easy to distinctly define all the functions of social support, since many of them do not have an apparent boundary (House, 1981; Thoits, 1995; Yeung & Fung, 2007). For instance, domestic helper could provide instrumental support to the elderly because of the dependence on them to attend to, while, on the other hand, older adults may regard this caring behavior as a kind of satisfaction of being accompanied. Thereby, during the process of differentiating those functional properties; it is indeed significant to take concrete analysis. In other words, to avoid any overlap in classification is not that likely.

From a holistic perspective, people in later life are assumed to pay more attention to their emotional needs because they may perceive the future time is limited (Carstensen, Isaacowitz, & Charles, 1999). Therefore, emotional and companionship support would be highly valued by older adults. Additionally, given that the current society has stepped into a developed information era, individuals have more accesses to different types of information than before. For the older adults living in the community, they may receive informational support from different types of social partners, from whom they can get useful information about insurance, allowance, and social welfare, that they are interested in. Research on informational support of the elderly in Chinese societies are lacking, thus the present study will examine the received informational support in relation with positive aging. Apart from this, receiving instrumental support is important in later life especially for the frail elderly. Then the study will take account of instrumental support and explore its resources and association with positive aging.

Other aspects of social support

For the dimension of social support, it has been demonstrated that social support does not only include the support offered by others, it also contains the behaviors of active help-seeking and helping others (Phillips, Siu, Yeh, & Cheng, 2008). And in many occasions, the degree of active help-seeking has a direct influence on the amount of support one receives. For reciprocal reasons, many capable older adults may attempt to help others in order to give others support if they are in need, such as working as volunteers, passing on the wisdom to their children or others, etc. In addition, the subjective appraisal such as the satisfaction of social support is also a reflection of the situation, which should also be taken into consideration in analysis.

2.3.3.3 Emotional Closeness

In social relations, individuals tend to have frequent and intimate interactions with a number of network members, which are considered as close social partners. Emotional closeness, which is derived from the relations with close social networks, was proposed. It was defined as a sense of concern, trust and caring for another individual and enjoyment of the relationship with that individual (Cunningham, 1986; Lee, Mancini, & Maxwell, 1990). Emotional closeness between two people leads them want to help each other, especially when the other is in need (Korchmaros & Kenny, 2001). So far, theories and research suggest that close relationship and social support are essential to the mental health of adults (Rathus & O'Leary, 1997). It is also found that strong emotional connections in later life predicted happiness and adjustment (Antonucci & Jakson, 1987). Emotional closeness to family and close friends was documented in the research of interrelationships. The interaction with familiar and close social partners was postulated to yield sufficient gains and benefits to individual's well-being in the premise of socioemotional selectivity theory (Carstensen, 1992). Empirical test also revealed that elderly people placed the greatest emphasis on the potential for affection categorizing prospective social partners (Fredrickson & Carstensen, 1990).

Previous research has found that emotional closeness is subject to the perceived psychological closeness between two persons, which might not be necessarily related with the rates of contacts and network size (Carstensen, 1992). Emotional closeness is displayed through the degrees of indicators showing willingness to interact or discuss intimate feelings with the other. As is documented in earlier sections, research on social network tends to focus on the quantity and quality of a network. In terms of emotional closeness in the present study, the research aims to explore one of the components of quantity—the breadth, and one of the components of quality—the depth. To be specifically, the breadth of emotional closeness points to the size of close social networks, that is, the number of close social partners. The depth of emotional closeness means the exchange of emotional support and intimacy in the interactions with close social partners. In the present study, emotional closeness of community-dwelling older adults in Hong Kong will be examined to look at its effect on positive aging. However, because there is a paucity of empirical studies of emotional closeness, especially among the elderly people, the current study will elaborate the operationalization of emotional closeness in later chapter.

2.4 Coping

2.4.1 Coping: Origins and Definitions

Coping is generally interpreted as a crucial variable influencing the adaptable outcomes of a person's struggle to get along or live well (Lazarus & DeLongis, 1983). Research on the topic of coping is usually related to people who are facing certain kinds of dilemmas or problems. In social gerontology, old people are experiencing a period of time when the physical and cognitive functions are not as good as before with their age increasing. And to some extent, the aging process could be understood as a kind of stress to the elderly under their circumstance. Therefore, the attitudes on how to cope with the aging process and remain adaptive to this distinct change have become a focus in research. The most significant point is that coping could well reveal the process of facing aging problems and how to conquer them, which is quite close to the original purpose of research—to study positive aging from a kind of dynamic view. Additionally, as to positive aging, coping strategies have been proved to play an important role in leading a healthy and positive sunset life of the elderly (Folkman, Lazarus, Pimley, & Novacek, 1987; Jonker, Comijs, Knipscheer, & Deeg, 2009).

There are many kinds of definitions and descriptions on coping. Folkman and Lazarus (1980) pointed out that there were two types of coping. The first type was problem-focused coping, aiming to solve problems or doing something to alter the source of the stress. The second was elaborated as emotional-focused coping, tending to alleviate or manage the emotional distress which was associated with the situation (Carver, Scheier, & Weintraub, 1989). This classification of coping sounds reasonable, it respectively addresses the two dimensions—behavioral direction and psychological direction. However, in terms of the distinction between the two, researchers have found other factors that respond to coping issue more than those two (as cited in Carver et al., 1989). In the coping process, there might not be apparent differentiation that people adopt methods to cope with stress from the problem-focused level or emotional-focused level. And it is argued that the process of many problem-focused coping behaviors usually involve emotion-focused facets (Thoits, 1995), so that this specific classification of coping may deserve further examination.

In addition, other researchers also have their own opinions on coping. Cohen and Lazarus (1979) provided four modes of coping, they were information seeking, direct action, inhibition of action, and intrapsychic. While it seems a bit difficult to clearly define these elements and meanwhile some paradox still exist among the modes of coping. So in the actual situations, it is really hard to differentiate coping behaviors by virtue of the four modes. Apart from that, other studies also reveal a series of factors contributing to coping behaviors. Religion (Koenig, George, & Siegler, 1988), activities to keep busy (Rosen, 1982), passive behaviors (Folkman et al., 1987), and education (Blanchard-Fields & Irion, 1988) have been proved to act as necessary roles in coping process.

As to this specific study, coping resources could be generally classified by virtue of the resource direction. It consists of a series of strategies from external or internal domains, which could be adopted to help solve concrete problems or expel stress. For the older adults, many of them try to seek for external resources to overcome troubles, for example, to look for actual help in curing diseases, improving cognitive abilities and increasing adaptation to environments. So these behaviors under the help of external sources are part of coping strategies. Additionally, due to the decline of physical health and personal competence, in many occasions old people could not totally handle the stress and problems. Then, psychological coping strategies are adopted by the elderly and utilized to cope with negative events. The two directions of coping resources could be encompassed in the following SOC model in successful aging (Baltes & Baltes, 1993).

2.4.2 *Coping Strategies: Dimensions and Theoretical Background*

2.4.2.1 Selection, Optimization and Compensation

Selection, optimization and compensation (SOC) is a kind of combined coping strategy containing both external and intrapersonal resources, which is widely used by the elderly people. It could be thought as the process of achieving the maximization of gains and minimization of losses (Ng, Woo, Kwan, & Chong, 2006). For the three essential elements of SOC, selection refers to restricting or reducing one's range of affairs and activities in the life world to fewer domains. This is because older people may have some declines in physical, psychological and other aspects, it requires to narrow down the life boundary and concentrate on those most important tasks. So in this adaptive process, a sense of selection from all the affairs with different priorities is crucial. The second component, optimization, goes to a kind of strategy which aims to maximize and augment the value or achievement during a person's whole life (Abraham & Hansson, 1995). It reflects a motivation to attain a better even best outcome of a chosen task. The proper utilization and development of one's competence is an important indicator of optimization. Lastly, compensation also results in a kind of adaptive potential or restriction when people encounter very big problems or high demands that they could not handle. Generally, compensation could be regarded as an active action to look for a pragmatic method to compensate for the loss, and it could be also classified as a coping strategy by virtue of external assistance. As to the elderly, a common interpretation of compensation skills is that they try to use presbyopic glasses, walking stick, and other instruments to help overcome difficulties. It is different from selection and optimization that when people try to use compensation skills, their current capacities are under the minimum standard of the tasks, whereas, to some extent, it is more flexible to adopt optimization and selection because it's depended on the individuals themselves.

The application of SOC in aging process is of great help, especially when the available resources and personal competence are very low. Previous study has demonstrated that older adults who usually use SOC could buffer the negative effect of restriction of resources and social reserves (Jopp & Smith, 2006). Additionally, it is also proved that older people who reported using the model of SOC in behaviors to manage aging life showed a respectively higher score on indicators such as subjective well-being and positive emotions in successful aging (Freund & Baltes, 1998).

2.4.2.2 The Meaning-Making Model of Coping

Park and Folkman (1997) have suggested a meaning-making model of coping to adjust to major traumas and loss. This meaning-making model of coping consists of global coping and situational coping, which are used to deal with situations of crisis, serious illness, disasters and severe accidents. Global meaning includes global beliefs and global goals. Global beliefs are the basic internal cognitive structures that individuals construct about the nature of the world. These structures guide people to construct the reality and interpret the global goals throughout life. Situational meaning points to the meaning that is formed in the interaction between a person's global meaning and the circumstances of a particular person-environment transaction. Situational meaning contains an initial appraisal of the meaning of the event and the process of searching for meaning, either or both of which may in turn affect global meaning (Park & Folkman, 1997). This stress-coping model provides explanations to coping strategies that are not problem-solving oriented. Past empirical studies revealed that making meaning following negative events was positively associated with adjustment to that event (Affleck, Tennen, & Gershman, 1985). However, other studies questioned whether it was helpful in adjustment (Clark, Henry, & Taylor, 1991) and found negative correlations between meaning-making and adjustment (Frazier & Schauben, 1994).

Noteworthy is that this model is beneficial to the elderly who are confronted with life crisis in later life. When people are confronted with difficult-to-solve problems, highly stressful events or traumatic loss, the problem-focused coping strategies might not be effective (Park, 2005). Research indicates that although some strategies of emotion-focused coping, for example, talking with others about the stressful experience, are helpful in coping, other emotion-focused strategies, such as avoidance, tend to be related to continued distress (Park, 2005).

The meaning-making model is a positive approach to cope with difficult-to-solve stressors through cognitive adaptation and make cosmic appraisals of the stress, lost or challenge occurred (Park & Folkman, 1997). It should be noticed that this model targets highly stressful life events and aims to achieve better psychological adjustment. For the current research, positive aging is a constructed concept with multi-facets, including physical and functioning health, CE, PE and life satisfaction. In line with the implications and empirical findings of this model, it was assumed that the meaning-making model could contribute to the subjective components of positive aging—life satisfaction, but was not necessarily associated with other four objective components. Furthermore, the current research did not aim to further examine the stressful events experienced by the older adults. Instead, it was expected to explore whether social support network could trigger certain type of coping strategy in contributing to positive aging, which is not the core of the model. In future research, the meaning-making model of coping is recommended to be examined in later life through a comprehensively designed study with explicit research questions related to global meaning and situational meaning. Moreover, the future sample should focus on people who have experienced highly stressful events.

2.4.2.3 The Three-Factor Model of Coping Strategies

Coping strategies represent the efforts, both behavioral and cognitive, that people invest in order to deal with stressful encounters (Lazarus & Folkman, 1984). These efforts would result in a variety of activities, classified according to problem-focused and emotion-focused modes, as differentiated by their function. Problem-focused coping contains various actions aimed at managing future danger or threat, expressed in the efforts the person invests in order to actually change his/her behaviors or interaction with the environment. In contrast, emotion-focused coping, is aimed at reducing, preventing or tolerating the emotional and bodily reactions that are characterized as stressful, which is more person oriented (Ben-Zur, 2002).

In the following research in terms of coping, various coping models and classification were developed. Endler and Parker (1990) suggested a third dimension, avoidance, which can include either task or person oriented strategies. Then Ben-Zur (1999) reported a construct of three factors: (1) problem/accommodation coping, composed of problem-focused strategies (e.g., active coping) as well as accommodation strategies (e.g., positive reinterpretation and growth); (2) emotion/support coping, composed of approach type of emotion-focused strategies (e.g., looking for social support for emotional reasons); and (3) avoidance/disengagement, composed of avoidance type of emotion-focused coping (e.g., denial).

Although the dyadic classification of coping, expressed as problem-focused coping and emotion-focused coping, has been widely adopted in a variety of empirical studies, the three coping meta-strategies proposed by Ben-Zur (1999) tend to be more comprehensively generalized. In addition to personal characteristics, age is another variable believed to have influence on dispositional coping strategies. They are exposed to a greater variety of events, negative as well as positive, and will have learned the effectiveness or otherwise of coping strategies they employed through the process of tackling problems as they arose during the course of their lives (Wong, 2003). Folkman et al. (1987) found that across a wide variety of life domains, young people use more active, problem-focused coping than older people, while the latter use more passive, emotion-focused coping than the former. Additionally, Blanchard-Fields, Jahnke, and Camp (1995) also have similar findings showing age differences in problem-solving style, with older adults (65–75 years) using less problem-focused coping and more passive-dependent coping than other age groups (e.g., adolescents, young or mid-age adults). To a large extent, previous findings addressed the specific feature of coping strategies in old age and pointed out the prevalence of passive/avoidance coping. Thus, taking into account of the characteristics of older adults, this study will take the three-factor model of coping into analysis—problem/accommodation coping (Problem); (2) emotion/support coping (Emotion); and (3) avoidance/disengagement (Avoidance).

2.4.2.4 Simplified Coping Strategies

Few previous literatures on coping had distinguished the differences between older adults and other age groups. Compared with adults, the old may experience less life events overall related with career and interrelationships because they've retired in society. And it was proclaimed that on the one hand the older people would experience more loss events, including those associated with declining health and loss of relatives and friends (Lazarus & DeLongis, 1983), on the other hand, at the same time, they were experiencing plenty of ordinary hassles of day-to-day living (Folkman et al., 1987), which was demonstrated to be strongly related with distress (Holahan, Holahan, & Belk, 1984). Thereby, the capacity of coping with the stressors and obstacles occurred in aging life seems pivotal in maintaining well-being.

From the previous studies, the well-known forms of problems-focused and emotion-focused coping strategies have been frequently explored (Carver et al., 1989). However, this classification had overlaps in terms of seeking social support, escape-avoidance, etc. As is discussed earlier, the SOC model was proposed to describe the adaptive process of aging as a specialization in particular functional domains as well as an adjustment to losses during later life (Lang & Carstensen, 1994). Nevertheless, this approach to study coping in late life might be too mechanical and instrumental for the old to carry out. And this coping model was originally proposed with the postulate that individuals were confronted with a lack of resources (e.g., opportunity for education, financial resource deficiency, and diseases), thus the application of SOC would assist to reach the goal. Since when people grow older, much of the declines (health and functioning) and losses (retirement and death) are no longer easy to regain, the less resources they possess might also stimulate fewer life goals. These changes would result in less initiative for the old in coping with problems and stressors. Therefore, the first two components in SOC model—selection and optimization, would probably not be often applied because they are more inclined to be pragmatic and problem-focused, which might go beyond the actual capacity of older people.

However, in accordance with some empirical studies, it was reported that a series of negative coping strategies (e.g., escape-avoidance, distancing) were proportionally more frequently used (Folkman et al., 1987). Additionally, researchers have examined a series of simplified coping strategies among Hong Kong older adults through a Simplified Coping Style Questionnaire (SCSQ) composed by active coping and passive coping, which acquired a satisfactory reliability (Cronbach alpha of the whole scale is 0.76) and construct validity (Wong, 2003). Thus, based on the actual situations that older people are facing with, the present study will adopt SCSQ as the tool to examine coping strategies among the community-dwelling older adults.

2.5 Domestic Helper as a Support Provider

2.5.1 Introduction

With the development of home-based or community-based care, more and more elderly people choose to stay at their familiar living environment to spend their sunset life. The traditional reliance on family caregivers to care for frail elders has undergone significant changes because of socioeconomic shifts, including the rise of the nuclear family and an increase in the female employment rate (Chong et al., in press). For the reason that the adult children are usually engaged with their own career and family, some new caregivers including domestic helpers, nurses, and home helpers, etc. emerge to provide informal care to older people (Lai, 2009). This trend is becoming more popular in mainland China and Hong Kong, as well as foreign countries such as Japan, Singapore, United States, etc. (Ekwall, Sivberg, & Hallberg, 2004; Lam et al., 2010; Ozeki, 1997; Wu, Carter, Goins, & Cheng, 2005; Yeoh & Huang, 2010). In mainland China, the full-time caregiver who works as an employee at home is normally named Baomu (nanny) (Wu et al. 2005), and their originalities vary from local citizens to migrant workers in rural areas. While in Hong Kong and some Asian countries, domestic helpers are mostly from Southeast Asia, for example, Philippines, Indonesia, and so on (Yeoh & Huang, 2010). In terms of caring for the elderly, domestic helpers are able to contribute to the implementation of long-term care. Furthermore, it is predicted that the role of domestic helpers would reveal its increasing importance in elderly care in the future (Zhai & Qiu, 2007).

2.5.2 The Hong Kong Context

As to Hong Kong, the importation of foreign domestic helpers (FDHs) has been authorized for many years since mid-1970s (Ozeki, 1997). FDH in Hong Kong is defined as someone who provides full-time, live-in domestic services at the employer's residence (Immigration Department, 2014). According to the statistics, the number of FDHs working in Hong Kong has reached to 21,500 in 1982, reflecting a substantial increase compared with 292,473 in 2011 (Ng, 2011). FDHs usually live in their employer's residence and assist with household chores (e.g., cooking and cleaning), escort elders to outdoor activities, and provide personal care on the basis of need. Generally, most of the FDHs are female. Hong Kong residents increasingly rely on FDHs to care for frail older people due to declines in extended family, busy lifestyles, and an increase in the female employment rate. Additionally, another reason is insufficient government-funded long-term care services, especially in terms of residential care (Chong et al., in press).

A review of the literature reveals that there are few studies focusing on the relationship between services provided by domestic helpers, service satisfaction and positive aging. Nevertheless, it seems that the elderly who live in their own home especially those who have bad cognitive and functional conditions incline to employ domestic helpers (Lam et al., 2010). From the angle of support provision, FDHs are usually able to provide regular instrumental support, since taking care of the elderly is considered as the duty for them. Apart from that, empirical studies also reveal that the support from domestic helpers could be beneficial to those who have suffered from diseases and disabilities, as a result, relieving the perception of pains and dementia symptoms of the older people (Kirby et al., 2000; Lam et al., 2010). Similarly, other researchers have found that the employment of a domestic helper predicted reduced institutionalization among the functionally and cognitively impaired elders (Chau et al., 2010). Additionally, it is also available that domestic helper acts as a companionship provider to their older employers. In daily routine, they accompany the older adults to walk around, to see the doctor or to go to community center. Thereby, this is another reason why older people have the demand of employing domestic helpers. For the relationship between both parties, the cultural diversities still dominate, and it is said that the Chinese are not willing to develop close or personal relationship with the Philippians beyond the necessary communication with them as employers (Ozeki, 1997). In addition, to a great degree, the established employ-employee relationship has restricted the further development of intimacy and equality. Then FDHs as elderly caregivers in Hong Kong are still unable to provide adequate and superior emotional support to the old.

2.5.3 *The Mainland China Context*

In mainland China, the situation of domestic helper seems a bit different. In several big cities of China, long-term care plan is carried out step by step in some experimental units of urban areas. In Shanghai, one of the highly industrialized and modernized cities, many kinds of social policies and social welfare services are created to facilitate caring for the elderly in community (Shanghai Civil Affairs Bureau, 2005). For those older adults who are prone to stay at home, there are several kinds of community services to choose. Baomu, which has the same meaning as domestic helper, is one of the dominant services provided to enhance the quality of life for the elderly. Other than Hong Kong, Baomu in Shanghai reflect more heterogeneity. Generally they do not have a unified originality; many of them are rural migrant women from the less developed provinces of China, while others are local laid-off workers (Wu et al., 2005). Migrant women tend to take full-time caring jobs as domestic Baomu, and those who reside locally are more prone to take part-time jobs. The role of Baomu is basically the same as those in Hong Kong. They provide various in-home services for elders and their families, helping them maintain the normal daily schedule. Additionally these basic services, Baomu are considered to provide social support to older people commonly, including accompanying elders in medical visits and provide escort (Wu et al., 2005). However, the employers of Baomu frequently reflect their hesitation in selecting an appropriate housemaid. For the rural migrant women, they cannot speak fluent local dialect, even Mandarin, leading to some communication problems when providing caring services to older adults. And it is also recorded that the stealing and cheating cases of migrant Baomu are more serious these years. So, it is not simple to employ a satisfactory and helpful Baomu in mainland China (Zhai & Qiu, 2007). Furthermore, there is rare positive evaluation towards the relationship between both parties. In general, the old employers could not totally accept their Baomu as close family members, and then receive little emotional support from them (Wu et al., 2005).

2.5.4 *FDHs in Singapore and Japan*

In foreign countries like Singapore, a dominant notion on elderly care is to put more emphasis on encouraging older people to live in their own homes (Aronson & Neysmith, 1996). In Hong Kong, migrant domestic helpers usually come from other countries and are regulated by a work permit system. Similarly, as the main instrumental support providers, domestic helpers in Singapore receive more kinds of education and training from the agency in case of later dissatisfaction from the elder employers (Yeoh & Huang, 2010). Adequate and appropriate training is very necessary and useful, which, to some extent, improve the quality of service. As to the relationship between both parties, it is suggested that the FDHs are caregivers who do not belong wholly to the public or private, they are not treated as rightful family members nor trained employees (Teo, 2006). This perspective is widely recognized as an effective way to reduce conflicts and friction in families.

For the situation in Japan, a Long-Term Care Insurance Scheme was implemented in 2000, reflecting a big reform and intensive attention to elderly care affairs (Campbell & Ikegami, 2000). This huge welfare program could be divided into community-based care and institutional care. In the agenda of community-based care, a distinguished measure consisting of home-helper services have been drawn much attention. For the home helpers, also known as home care workers, is a new occupation created by the Japanese government. People who work as home helpers must receive professional training and finally attain a certificate before starting their jobs. The general tasks of home-helper mainly include providing care such as assistance with bathing, toileting, feeding, etc. to patients who have difficulties in daily living because of physical and mental disorder (Murashima, Nagata, Magilvy, Fukui, & Kayama, 2002). These home helpers are generally regulated by relevant social welfare organizations and paid by the Long-Term Care Insurance. After the launch of this scheme, the number of home helpers is increasing very rapidly. And this approach to elderly care has intended to help older adults to live more independently to a great extent. Additionally, the appearance of home helpers also benefits the normal caregivers in families by relieving some of their stress (Traphagan, 2003). In past empirical studies, it is argued that the home-helper program forms a compensatory elder care system which would augment family-provided care and social support (Traphagan, 2003). Specifically, in line with the essence of Japan's Long-Term Care Insurance Scheme, the wide promotion of home helpers have improved the service quality on elderly caregiving, which is strongly supported by the masses. It is claimed that the new program which departs from past Japanese practices shifts the responsibility for caregivers from the family to the state (Campbell & Ikegami, 2000), bringing about more choices and opportunities in elderly care services. So it could be noticed that home-helpers are becoming a popular caring approach to promote Long-Term Care for the elderly, which provides some inspiration and reference for Chinese societies.

2.5.5 Summary

From the relevant literature and policy report, it is apparent that the large amounts of domestic helpers play an indispensable role in facilitating community-based and home-based care services. However, the mutual relationship of employer-domestic helper, quality of caregiving, and satisfaction in both parties still need further exploration. In the current study, the exploration towards FDHs who acts as support providers would be given a priority in research, aiming to excavate more significant findings to react to the current situation of caregiving to frail elders in Hong Kong.

References

Abraham, J. D., & Hansson, R. O. (1995). Successful aging at work: An applied study of selection, optimization, and compensation through impression management. *The Journals of Gerontology, Series B: Psychological Sciences and Social Sciences, 50*(2), 94–103.

Adler, P. S., & Kwon, S. (2002). Social capital: Prospects for a new concept. *The Academy of Management Review, 27*(1), 17–40.

Affleck, G., Tennen, H., & Gershman, K. (1985). Cognitive adaptation to high-risk infants: The search for mastery, meaning, and protection from future harm. *American Journal of Mental Deficiency, 89*(6), 653–656.

Antonucci, T. C., & Jackson, J. (1987). Social support, interpersonal efficacy, and health: A life course perspective. In L. L. Carstensen & B. A. Edelstein (Eds.), *Handbook of clinical gerontology* (pp. 291–311). Elmsford, NY: Pergamon Press.

Aronson, J., & Neysmith, S. M. (1996). You're not just in there to do the work. *Gender & Society, 10*(1), 59–77.

Atchley, R. C. (2006). Continuity, spiritual growth, and coping in later adulthood. *Journal of Religion, Spirituality & Aging, 18*(2–3), 19–29.

Baltes, P. B., & Baltes, M. M. (1990). Psychological perspectives on successful aging: The model of selective optimization with compensation. In P. B. Baltes & M. M. Baltes (Eds.), *Successful aging: Perspectives from the behavioral sciences* (pp. 1–34). New York: Cambridge University.

Baltes, M. M., & Carstensen, L. L. (1996). The process of successful ageing. *Ageing & Society, 16*(4), 397–422.

Barrera, M., & Ainlay, S. L. (1983). The structure of social support: A conceptual and empirical analysis. *Journal of Community Psychology, 11*(2), 133–143.

Ben-Zur, H. (1999). The effectiveness of coping meta-strategies: Perceived efficiency, emotional correlates and cognitive performance. *Personality and Individual Differences, 26*(5), 923–939.

Ben-Zur, H. (2002). Coping, affect and aging: The roles of mastery and self-esteem. *Personality and Individual Differences, 32*(2), 357–372.

Blanchard-Fields, F., & Irion, J. C. (1988). Coping strategies from the perspective of two developmental markers: Age and social reasoning. *The Journal of Genetic Psychology, 149*(2), 141–151.

Blanchard-Fields, F., Jahnke, H. C., & Camp, C. (1995). Age differences in problem-solving style: the role of emotional salience. *Psychology and Aging, 10*(2), 173–180.

Blazer, D. G. (1982). Social support and mortality in an elderly community population. *American Journal of Epidemiology, 115*(5), 684–694.

Bourdieu, P. (1985). The forms of capital. In J. G. Richardson (Ed.), *Handbook of theory and research for the sociology of education*. New York: Greenwood.

Bowling, A. (1993). The concepts of successful and positive aging. *Family Practice, 10*(4), 449–453.

Bowling, A., & Dieppe, P. (2005). What is successful ageing and who should define it? *British Medical Journal, 331*(7531), 1548–1551.

Bowling, A., & Gabriel, Z. (2004). An integrational model of quality of life in older age. Results from the ESRC/MRC HSRC quality of life survey in Britain. *Social Indicators Research, 69*(1), 1–36.

Bowling, A., & Iliffe, S. (2006). Which model of successful ageing should be used? Baseline findings from a British longitudinal survey of ageing. *Age and Ageing, 35*(6), 607–614.

Bradburn, N. M. (1969). *The structure of psycho-logical well-being*. Chicago: Aldine.

Campbell, A. (1981). *The sense of well-being in America*. New York: McGraw-Hill.

Campbell, J. C., & Ikegami, N. (2000). Long-term care insurance comes to Japan. *Health Affairs, 19*(3), 26–39.

Caplan,G., & Killilea, M. (1976). *Support systems and mutual help: Multidisciplinary explorations*. Grune & Stratton: New York.

Carstensen, L. L. (1992). Social and emotional patterns in adulthood: support for socioemotional selectivity theory. *Psychology and Aging, 7*(3), 331–338.

Carstensen, L. L., Isaacowitz, D. M., & Charles, S. T. (1999). Taking time seriously: A theory of socioemotional selectivity. *American Psychologist, 54*(3), 165–181.
Carver, C. S., Scheier, M. F., & Weintraub, J. K. (1989). Assessing coping strategies: A theoretically based approach. *Journal of Personality and Social Psychology, 56*(2), 267–283.
Chau, P. H., Kwok, T., Woo, J., Chan, F., Hui, E., & Chan, K. C. (2010). Disagreement in preference for residential care between family caregivers and elders is greater among cognitively impaired elders group than cognitively intact elders group. *International Journal of Geriatric Psychiatry, 25*(1), 46–54.
Chen, X., & Silverstein, M. (2000). Intergenerational social support and the psychological well-being of older parents in China. *Research on Aging, 22*(1), 43–65.
Cheng, S. T., Lee, C. K. L., Chan, A., Leung, E. M. F., & Lee, J. J. (2009). Social network types and subjective well-being in Chinese older adults. *The Journals of Gerontology, Series B: Psychological Sciences and Social Sciences, 64*(6), 713–722.
Cheung, J., Kwan, Y. H., Ng, S. H., Ngan, R., Lau, A., Leung, M. F., ..., Chan, K. K. (2002). Conceptions of positive aging among older people in Hong Kong. In Kwan, Y. H. (ed.), *Aging Hong Kong: Issues facing an aging society* (pp. 12–39). Hong Kong: Cosmos.
Chong, A. M. L, & Ng, G. F. M. (2010). Bridging social capital—Theories and practice. In Ng, Cheung and Prakash (Eds.) *Social Capital in Hong Kong* (pp. 75–91). Hong Kong: City University of Hong Kong Press.
Chong, A. M. L., Ng, S. H., Woo, J., & Kwan, A. Y. H. (2006). Positive ageing: The views of middle-aged and older adults in Hong Kong. *Ageing & Society, 26*(2), 243–265.
Clark, L. F., Henry, S. M., & Taylor, D. M. (1991). Cognitive examination of motivation for child-bearing as a factor in adjustment to infertility. In A. L. Stanton & C. Dunkel-Schetter (Eds.), *Infertility: Perspectives from stress and coping research* (pp. 157–180). New York: Plenum.
Cobb, S. (1976). Social support as a moderator life stress. *Psychosomatic Medicine, 39*(5), 300–314.
Cobb, S. (1979). Social support and health through the life course. In M. W. Riley (Ed.), *Aging from birth to death: Interdisciplinary perspective* (pp .93–106). Boulder, Co: Westview Press.
Cohen, F., & Lazarus, R. S. (1979). Coping with the stresses of illness. In G. S. Stone, F. Cohen, & N. E. Adler (Eds.), *Health Psychology: A Handbook* (pp. 77–112). San Francisco: Jossey-Bass.
Cohen, S., & McKay, G. (1984). Social support, stress, and the buffering hypothesis: A theoretical analysis. *Handbook of Psychology and Health, 4,* 253–267.
Cohen, S. E., & Syme, S. (1985). *Social support and health.* Cambridge: Academic Press.
Coleman, J. S. (1994). *Foundations of social theory.* Cambridge: Belknap Press.
Cresci, M. K. (2001). The relationship between informational support and post-injury functional status in older women recovering from a hip fracture. *Educational Gerontology, 27*(3), 281–295.
Crowther, M. R., Parker, M. W., Achenbaum, W. A., Larimore, W. L., & Koenig, H. G. (2002). Rowe and Kahn's model of successful aging re-visited: Positive spirituality—the forgotten factor. *The Gerontologist, 42*(5), 613–620.
Cunningham, M. R. (1986). Levites and brother's keepers: A sociobiological perspective on prosocial behavior. *Humbolt Journal of Social Relations, 13*(1/2), 35–67.
Diener, E. (1984). Subjective well-being. *Psychological Bulletin, 95*(3), 542–575.
Diener, E., Suh, E. M., Lucas, R. E., & Smith, H. L. (1999). Subjective well-being: Three decades of progress. *Psychological Bulletin, 125*(2), 276–302.
Ekwall, A., Sivberg, B., & Hallberg, I. R. (2004). Dimensions of informal care and quality of life among elderly family caregivers. *Scandinavian Journal of Caring Sciences, 18*(3), 239–248.
Endler, N. S., & Parker, J. D. (1990). Multidimensional assessment of coping: A critical evaluation. *Journal of Personality and Social Psychology, 58*(5), 844–854.
Erikson, E. H. (1959). *Identity and the life cycle: Selected papers.* Madison, CT: International Universities Press.
Fernández-Ballesteros, R. (2010). Quality of life in old age: Problematic issues. *Applied Research in Quality of Life, 6*(1), 1–20.
Fernandez-Ballesteros, R. (2011). Positive ageing: Objective, subjective, and combined outcomes. *Electronic Journal of Applied Psychology, 7*(1), 22–30.

Festinger, L. (1954). A theory of social comparison process. *Human Relations, 7*(2), 117–140.
Folkman, S., & Lazarus, R. S. (1980). An analysis of coping in a middle-aged community sample. *Journal of Health and Social Behavior, 21*(3), 219–239.
Folkman, S., Lazarus, R. S., Pimley, S., & Novacek, J. (1987). Age differences in stress and coping processes. *Psychology and Aging, 2*(2), 171–184.
Fozard, J. L., & Popkin, S. J. (1978). Optimizing adult development: Ends and means of an applied psychology of aging. *American Psychologist, 33*(11), 975–989.
Frazier, P. A., & Schauben, L. J. (1994). Causal attributions and recovery from rape and other stressful life events. *Journal of Social and Clinical Psychology, 13*(1), 1–14.
Fredrickson, B. L., & Carstensen, L. L. (1990). Choosing social partners: How old age and anticipated endings make people more selective. *Psychology and Aging, 5*(3), 335–347.
Freeman, L. C. (2004). *The development of social network analysis*. Vancouver: Empirical Press.
Freund, A. M., & Baltes, P. B. (1998). Selection, optimization, and compensation as strategies of life management: Correlations with subjective indicators of successful aging. *Psychology and Aging, 13*(4), 531–543.
George, L. K. (1979). The happiness syndrome: Methodological and substantive issues in the study of social-psychological well-being in adulthood. *The Gerontologist, 19*(2), 210–216.
Gergen, M. M., & Gergen, K. J. (2001). Positive aging: new images for a new age. *Ageing International, 27*(1), 3–23.
Hill, R. D. (2005). *Positive aging: A guide for mental health professionals and consumers*. New York: W. W. Norton & Co Inc.
Hill, R. A., & Dunbar, R. I. M. (2003). Social network size in humans. *Human Nature, 14*(1), 53–72.
Holahan, C. K., Holahan, C. J., & Belk, S. S. (1984). Adjustment in aging: The roles of life stress, hassles, and self-efficacy. *Health Psychology, 3*(4), 315–328.
Holmén, K., & Furukawa, H. (2002). Loneliness, health and social network among elderly people—a follow-up study. *Archives of Gerontology and Geriatrics, 35*(3), 261–274.
Homans, G. C. (1961). *Social behavior*. NY: Harcourt Brace.
House, J. S. (1981). *Work stress and social support*. MA: Addison-Wesley Longman.
House, J. S., Umberson, D., & Landis, K. R. (1988). Structures and processes of social support. *Annual Review of Sociology, 14*(1), 293–318.
Immigration Department. (2014). *Frequently asked questions (FAQs): Foreign domestic helpers*. Hong Kong, China: Government of the Hong Kong Special Administrative Region. Retrieved from http://www.immd.gov.hk/en/faq/foreign-domestic-helpers.html.
Ingersoll-Dayton, B., Morgan, D., & Antonucci, T. (1997). The effects of positive and negative social exchanges on aging adults. *The Journals of Gerontology, Series B: Psychological Sciences and Social Sciences, 52*(4), 190–199.
Jonker, A. A. G. C., Comijs, H. C., Knipscheer, K. C. P. M., & Deeg, D. J. H. (2009). The role of coping resources on change in well-being during persistent health decline. *Journal of Aging and Health, 21*(8), 1063–1082.
Jopp, D., & Smith, J. (2006). Resources and life-management strategies as determinants of successful aging: On the protective effect of selection, optimization, and compensation. *Psychology and Aging, 21*(2), 253–265.
Kahn, R. L., & Antonucci, T. C. (1980). Convoys over the life course: attachment, roles and social support. In P. B. Baltes & O. Brim (Eds.), *Life-span development and behavior* (Vol. 4, pp. 253–286). New York: Academic Press.
Kirby, M., Denihan, A., Bruce, I., Coakley, D., & Lawlor, B. A. (2000). The pattern of support networks among the community dwelling elderly in urban Ireland: Variations with mental disorder. *Irish Journal of Psychological Medicine, 17*(2), 43–49.
Koenig, H. G., George, L. K., & Siegler, I. C. (1988). The use of religion and other emotion-regulating coping strategies among older adults. *The Gerontologist, 28*(3), 303–310.
Korchmaros, J. D., & Kenny, D. A. (2001). Emotional closeness as a mediator of the effect of genetic relatedness on altruism. *Psychological Science, 12*(3), 262–265.

Lai, W. L. (2009). From burden to depressive symptoms: The case of Chinese-Canadian family caregivers for the elderly. *Social Work Health Care, 48*(4), 432–449.
Lam, L. C., Lee, J. S., Chung, J. C., Lau, A., Woo, J., & Kwok, T. C. (2010). A randomized controlled trial to examine the effectiveness of case management model for community dwelling older persons with mild dementia in Hong Kong. *International Journal of Geriatric Psychiatry, 25*(4), 395–402.
Lang, F. R., & Carstensen, L. L. (1994). Close emotional relationships in late life: Further support for proactive aging in the social domain. *Psychology and Aging, 9*(2), 315–324.
Lau, A., Chi, I., & McKenna, K. (1998). Self-perceived quality of life of Chinese elderly people in Hong Kong. *Occupational Therapy International, 5*(2), 118–139.
Lawton, M. P. (1983). Environment and other determinants of well-being in older people. *The Gerontologist, 23*(4), 349–357.
Lazarus, R. S., & DeLongis, A. (1983). Psychological stress and coping in aging. *American Psychologist, 38*(3), 245–254.
Lazarus, R. S., & Folkman, S. (1984). *Stress appraisal, and coping*. New York: Springer.
Lee, T. R., Mancini, J. A., & Maxwell, J. W. (1990). Sibling relationships in adulthood: Contact patterns and motivations. *Journal of Marriage and the Family, 52*(2), 431–440.
Lee, R. P. L., Ruan, D., & Lai, G. (2005). Social structure and support networks in Beijing and Hong Kong. *Social Networks, 27*(3), 249–274.
Li, C., Wu, W., Jin, H., Zhang, X., Xue, H., He, Y., …, Zhang, M. (2006). Successful aging in Shanghai, China: Definition, distribution and related factors. *International Psychogeriatrics, 18*(3), 551–563.
Liang, J. (1985). A structural integration of the affect balance scale and the life satisfaction Index A. *Journal of Gerontology, 40*(4), 552–561.
Lin, N., Simeon, R. S., Ensel, W. M., & Kuo, W. (1979). Social support, stressful life events, and illness: A model and an empirical test. *Journal of Health and Social Behavior*, *20*(2), 108–119.
Lykken, D., & Tellegen, A. (1996). Happiness is a stochastic phenomenon. *Psychological Science, 7*(3), 186–189.
Michalos, A. C. (1985). Multiple discrepancies theory (MDT). *Social Indicators Research, 16*(4), 347–413.
Murashima, S., Nagata, S., Magilvy, J. K., Fukui, S., & Kayama, M. (2002). Home care nursing in Japan: A challenge for providing good care at home. *Public Health Nursing, 19*(2), 94–103.
Naughton, M. J., & Wiklund, I. (1993). A critical review of dimension-specific measures of health-related quality of life in cross-cultural research. *Quality of Life Research, 2*(6), 397–432.
Neugarten, B. L., Havighurst, R. J., & Tobin, S. S. (1961). The measurement of life satisfaction. *Journal of Gerontology, 16*(2), 134–143.
Ng, C. N. (2011, July 11). Foreign domestic helpers appeal for right of abode in Hong Kong. *Sing Tao Daily*, A2.
Ng, S. H., Cheung, C. K. J., Chong, A. M. L., Woo, J., Kwan, A. Y. H., & Lai, S. F. Y. (2011). Aging well socially through engagement with life: adapting Rowe and Kahn's model of successful aging to Chinese cultural context. *The International Journal of Aging and Human Development, 73*(4), 313–330.
Ng, S. H., Woo, J., Kwan, A., & Chong, A. (2006). Positive ageing: A conceptual framework. *Korean Journal of Psychological and Social Issues, 12*(5), 29–43.
Okun, M. A. (1995). Subjective well-being. In George L. Maddox (Ed.), *The encyclopedia of aging: A comprehensive resource in gerontology and geriatrics* (2nd ed., pp. 909–912). New York: Springer Publishing Company.
Okun, M. A., & Stock, W. A. (1987). Correlates and components of subjective well-being among the elderly. *Journal of Applied Gerontology, 6*(1), 95–112.
Otte, E., & Rousseau, R. (2002). Social network analysis: A powerful strategy, also for the information sciences. *Journal of Information Science, 28*(6), 441–453.
Ozeki, E. (1997). Migration frontier of Filipino women: Ethnic relations of Filipina domestic helpers with Chinese employers in Hong Kong. *South East Asian Studies, 34*, 676–695.

Palys, T. S., & Little, B. R. (1983). Perceived life satisfaction and the organization of personal project systems. *Journal of Personality and Social Psychology, 44*(6), 1221–1230.
Pang-White, A. A. (2009). Reconstructing modern ethics: Confucian care ethics. *Journal of Chinese Philosophy, 36*(2), 210–227.
Park, C. L. (2005). Religion as a meaning-making framework in coping with life stress. *Journal of Social Issues, 61*(4), 707–729.
Park, C. L., & Folkman, S. (1997). Meaning in the context of stress and coping. *Review of General Psychology, 1*(1), 115–144.
Phillips, D. R., Siu, O. L., Yeh, A. G. O., & Cheng, K. H. C. (2008). Informal social support and older persons' psychological well-being in Hong Kong. *Journal of Cross-Cultural Gerontology, 23,* 39–55.
Pinquart, M., & Sorensen, S. (2000). Influences of socioeconomic status, social network, and competence on subjective well-being in later life: A meta-analysis. *Psychology and Aging, 15,* 187–224.
Pruchno, R. A., Wilson-Genderson, M., & Cartwright, F. (2010). A two-factor model of successful aging. *The Journals of Gerontology Series B, 65*(6), 671–679.
Putnam, R. D. (2001). *Bowling alone: The collapse and revival of american community*. New York: Simon and Schuster.
Rathus, J. H., & O'Leary, D. (1997). Spouse-specific dependency scale: Scale development. *Journal of Family Violence, 12*(2), 159–168.
Reichstadt, J., Depp, C. A., Palinkas, L. A., Folsom, D. P., & Jeste, D. V. (2007). Building blocks of successful aging: A focus group study of older adults' perceived contributors to successful aging. *American Journal of Geriatric Psychology, 15*(3), 194–201.
Robison, L. J., Schmid, A. A., & Siles, M. E. (2002). Is social capital really capital? *Review of Social Economy, 60*(1), 1–21.
Rosen, C. (1982). Ethnic differences among impoverished rural elderly in use of religion as a coping mechanism. *Rural Community Psychology, 3*(2), 27–34.
Rowe, J. W., & Kahn, R. L. (1987). Human aging: Usual and successful. *Science, 237*(4811), 143.
Rowe, J. W., & Kahn, R. L. (1997). Successful aging. *The Gerontologist, 37*(4), 433–440.
Rowe, J. W., & Kahn, R. L. (1998). *Successful aging*. New York: Pantheon.
Russell, D. W., & Cutrona, C. E. (1991). Social support, stress, and depressive symptoms among the elderly: Test of a process model. *Psychology and Aging, 6*(2), 190–201.
Ryff, C. D. (1982). Successful aging: A developmental approach. *The Gerontologist, 22*(2), 209–214.
Sampson, E. E. (1981). Cognitive psychology as ideology. *The American Psychologist, 36*(7), 730–743.
Schwarz, N., & Strack, F. (1999). Reports of subjective well-being: Judgmental processes and their methodological implications. In D. Kahneman, E. Diener, & N. Schwarz (Eds.), *Well-being: The foundations of hedonic psychology*. New York: Russell Sage Foundation.
Schwarzer, R., & Leppin, A. (1991). Social support and health: A theoretical and empirical overview. *Journal of Social and Personal Relationships, 8*(1), 99–127.
Scott, J. (2007). *Social network analysis: A handbook*. CA: Sage.
Shanghai Civil Affairs Bureau. (2005). Framework to guide the development of China charity (2006–2010). Retrieved online from http://www.shmzj.gov.cn/gb/shmzj/node8/node787/node887/userobject1ai23797.html.
Shinn, M., Lehmann, S., & Wong, N. W. (1984). Social interaction and social support. *Journal of Social Issues, 40*(4), 55–76.
Shumaker, S. A., & Brownell, A. (1983). A taxonomy of social support: Disentangling a conceptual morass. In A. Brownell & S. A. Shumaker (Chairs), *Annual Meeting of the American Psychological Association, Anaheim*, CA.
Son, J., & Lin, N., (2006, August). *Searching for a generalizable social support structure: Comparison between the U.S. and Taiwan.* Paper presented at the annual meeting of the American Sociological Association, Montreal Convention Center, Montreal, Quebec, *Canada Online* <PDF>. 2009-05-25 from http://www.allacademic.com/meta/p102965_index.html.

Stock, W. A., Okun, M. A., & Benin, M. (1986). Structure of subjective well-being among the elderly. *Psychology and Aging, 1*(2), 91–102.
Strawbridge, W. J., Cohen, R. D., Shema, S. J., & Kaplan, G. A. (1996). Successful aging: Predictors and associated activities. *American Journal of Epidemiology, 144*(2), 135–141.
Strawbridge, W. J., Wallhagen, M. I., & Cohen, R. D. (2002). Successful aging and well-being: Self-rated compared with Rowe and Kahn. *The Gerontologist, 42*(6), 727–733.
Tan, J., Ward, L., & Ziaian, T. (2011). Comparing definitions of successful ageing: The case of Anglo- and Chinese-Australians. *Electronic Journal of Applied Psychology, 7*(1), 15–21.
The Shorter Oxford English Dictionary. (1990). Oxford: Oxford University Press.
Thoits, P. A. (1995). Stress, coping, and social support processes: Where are we? What next? *Journal of Health and Social Behavior, 35*(Extra Issue), 53–79.
Ting, W. F. (2010). The praxis of social capital. In Ng, Cheung and Prakash (Eds.) *Social capital in Hong Kong* (pp. 37–70). Hong Kong: City University of Hong Kong Press.
Traphagan, J. W. (2003). Independence, security, and the intergenerational social contract: Home-helper services and elder care in rural Japan. *Care Management Journals, 4*(4), 216–222.
Van Tilburg, T. G. (1995). Delineation of the social network and differences in network size. In C. P. M. Knipscheer, J. de Jong Gierveld, T. G. van Tilburg, & P. A. Dykstra (Eds.), *Living arrangements and social networks of older adults* (pp. 83–96). Amsterdam: VU University Press.
Vandervoort, D. (1999). Quality of social support in mental and physical health. *Current Psychology, 18*(2), 205–222.
Vaux, A. (1988). *Social support: Theory, research, and intervention*. New York: Praeger.
Victor, C. R., Scambler, S. J., Bowling, A., & Bond, J. (2005). The prevalence of, and risk factors for, loneliness in later life: A survey of older people in Great Britain. *Ageing & Society, 25*(3), 357–375.
Wan, C. K., Jaccard, J., & Ramey, S. L. (1996). The relationship between social support and life satisfaction as a function of family structure. *Journal of Marriage and Family, 58*(2), 502–513.
Wellman, B. (1981). Applying social network analysis to the study of social support. In B. H. Gottlieb (Ed.), *Social networks and social support* (pp. 171–200). Beverly Hills: SAGE Publications.
Wenger, G. C. (1984). *The supportive network: Coping with old age*. London: Allen & Unwin.
Wills, T. A. (1985). Supportive functions of interpersonal relationships. In S. Cohen & S. L. Syme (Eds.), *Social support and health* (pp. 61–82). Orlando, Florida: Academic Press INC.
Wilson, W. R. (1967). Correlates of avowed happiness. *Psychological Bulletin, 67*(4), 294–306.
Wong, B. (2003). *The moderating effect of coping on depression in the elderly: A Hong Kong study*. Unpublished M. Phil. thesis, The University of Hong Kong.
World Health Organization. (1993). *Measuring quality of life: The development of the world health organization quality of life instrument (WHOQOL)*. Geneva: WHO.
Wu, B., Carter, M. W., Goins, R. T., & Cheng, C. (2005). Emerging services for community-based long-term care in urban China: A systematic analysis of Shanghai's community-based agencies. *Journal of Aging and Social Policy, 17*(4), 37–60.
Yeoh, B. S., & Huang, S. (2010). Foreign domestic workers and home-based care for elders in Singapore. *Journal of Aging & Social Policy, 22*(1), 69–88.
Yeung, G. T. Y., & Fung, H. H. (2007). Social support and life satisfaction among Hong Kong Chinese older adults: Family first? *European Journal of Ageing, 4*(4), 219–227.
Young, Y., Frick, K. D., & Phelan, E. A. (2009). Can successful aging and chronic illness coexist in the same individual? A multidimensional concept of successful aging. *Journal of the American Medical Directors Association, 10*(2), 87–92.
Zhai, X., & Qiu, R. Z. (2007). Perceptions of long-term care, autonomy, and dignity, by residents, family and caregivers: The Beijing experience. *Journal of Medicine and Philosophy, 32*(5), 425–445.

Chapter 3
Conceptual Framework

This chapter explicates the conceptual framework of the current research, on the basis of the previous background chapters. According to the research purpose, the theoretical background of the whole study, the relevant operational definitions, and hypotheses are described. Finally, the conceptual framework is generalized to facilitate the research design.

3.1 Theoretical Background of the Study

In social gerontology, scholars have created specific theories or approaches to uncover the questions or issues that are brought by the ageing process. A series of perspectives should be acknowledged. On the one hand, a number of theories focus on the role in later life solely. Activity theory is a theoretical approach that encourages the elderly to take part in social activities and have more social interaction, which is beneficial to the process of coping with aging as well as improving well-being and emotional adjustment (Hill, 2005; Lynott & Lynott, 1996). Afterwards, in contrast to activity theory, disengagement theory was raised (Cumming, 1961). It emphasized that less involvement of social roles were common in older people's normal aging progress. On the other hand, scholars tend to use a life span perspective to understand aging. Theories of late adulthood have been employed in research, in conjunction with the life span insights in younger times. For instance, the eighth stage of Erikson's theory of human life circle claimed that in midlife, men and women regarded aging as a positive process to grow beyond the rational ego-centered perspective of adulthood (Erikson, 1950). Additionally, the theory of gero-transcedence depicted the final stage in a natural process, which moved toward wisdom and maturation. The gero-transcendent individual may experience a new feeling of cosmic communication with the spirit of the universe, a redefinition of time, space, life and death, and a redefinition of the self (Tornstam, 2000).

S. Liu, *Social Support Networks, Coping and Positive Aging Among the Community-Dwelling Elderly in Hong Kong*,
https://doi.org/10.1007/978-981-13-3783-3_3

In regard to the current research, the cross-sectional design restricts the exploration from a life span perspective. However, the study was guided by some implications of the life span perspective to understand research questions. It should be noted that the present study could not test the hypotheses that were related to life trajectory, but was able to exhibit the situation in late adulthood by virtue of the selected life span theories at a single point of time. In this section, the theories and theoretical model that guide the current study are presented. The convoy model of social support is introduced, followed by the socioemotional selectivity theory. Then the hierarchical compensatory model and strengths perspective are outlined.

3.1.1 Convoy Model of Social Support

As stated in the previous chapter, one of the research objectives is to examine the structural and functional properties of social support network amongst the older adults, which are taken as the main independent variables. In reviewing the past literature related to social support and social network, the *Convoy Model of Social Support* is proposed to serve as one of the theoretical bases in this study.

Social support convoy is a model first proposed by Kahn and Antonucci (1980), describing a set of roles providing supportive interactions which are distributed into different layers across one's life span. Initially, the convoy was suggested coming from a personal network of family, friends, and others—a network that cut across the role boundaries in the interpersonal relationships and interactions (Kahn & Antonucci, 1980). In the convoy model, individuals are thought to have a dynamic system of relationships that envelop them over a life time (Depner & Ingersoll-Dayton, 1988). People have interactions with their convoy members by giving and receiving social support (e.g., emotional support, instrumental support) so as to assist in meeting their needs to maintain well-being and cope with stressors in life. From the lifespan perspective, it is claimed that as people age the structural and functional characteristics among convoy members could vary in a normative manner with life-cycle stage and the perception of closeness in a meaningful and predictable manner (Antonucci & Akiyama, 1987b).

In terms of the structural convoy of older adults, research found that close family members mostly made up the inner circle with more exchange social support with each other; nevertheless, they were prone to provide less support across specific relationships, including children, siblings and friends (Depner & Ingersoll-Dayton, 1988). Additionally, results showed that older people might own a smaller network than younger adults because the older were hypothesized to have a reduced number of convoy members with the losses through death or other external life circumstances (Antonucci & Akiyama, 1987b). Previous studies also indicated that although the role of family members and intimate members were significant and close in most elder's convoy, friends, neighbors who were considered as people with whom they enjoy spending time, engage in leisure activities and attain psychological well-being were noticeable in the convoy memberships (Antonucci and Akiyama, 1995).

For the functional components of social convoy among older adults, it has been demonstrated that the focal person in the convoy model might provide and receive less emotional support to their children, while it is widely believed that emotional support from adult children is a very important resource exchange which does good to aging adjustment of the older people (Antonucci and Akiyama, 1995; Depner & Ingersoll-Dayton, 1988). Evidence revealed that the lower number of siblings in the convoys of older people was associated with the decreased exchange of emotional and health-related support (Depner & Ingersoll-Dayton, 1988). Additionally, it is reported that respondents would receive more various types of support from the inner-circle members than the middle- or outer-circle members (Antonucci & Akiyama, 1987b). Remarkably, when compared with the situation of youth and adult groups from a lifespan perspective, the older groups were generally found to provide and receive less social support, which could be interpreted that their declined health status and less involvement in workplace decreased willingness to invest to relationships (Berges, Dallo, DiNuzzo, Lackan, & Weller, 2006; Carstensen, 1992; Levitt, Weber, & Guacci, 1993).

Some arguments on the convoy model of social support derived from the empirical studies are inconsistent with the original theory. Regarding the size of support network, research shows that there is not an obvious minimization of convoy size of the older people (Antonucci & Akiyama, 1987b; Depner & Ingersoll-Dayton, 1988). In addition to this, it was claimed that women inclined to be involved in more exchange of social support with their convoy members since women were thought to maintain more close supportive relationships than men (Antonucci & Akiyama, 1987a), however, statistics revealed that there was no increase in the percentage of women in the support networks of the oldest informants (Antonucci & Akiyama, 1987b).

To conclude, the convoy of social support is an integrated model to describe the structural, functional properties, and changes of social support provision and receipt among the focal person and the convoy members from a life course perspective. The layer changes of convoy members and relationships reflect that individuals move through their lifetimes surrounded by people who are close and important to them and who have a significant influence on their life and well-being (Antonucci & Akiyama, 1995). To facilitate the older people's sunset life, the vital role of family and friends should be underscored for their crucial effect on elder's well-being throughout the entire life course. Additionally, the subjective satisfaction towards the quality of social support needs much attention because of its close association with older adults' health and psychological well-being. In a word, to further maintain and optimize the social relationships in the convoy model in old age could potentially offer useful vehicles for achieving a positive and successful ageing life (Antonucci & Akiyama, 1995).

By adopting the theoretical approach of convoy model, the present study explored the situation of support providers who pertained to the three layers in an older person's convoy. However, it should be noticed that the convoy model is a life-span approach which examines the trajectory of social support across one's life. Therefore, the current study is not able to explore the changes of convoy because of the limitation of the cross-sectional research design. Consequently, the inevitable drawback of the present study is that it could only examine the facets of the convoy model at a single point of time. Therefore, to attain more information regarding the changes of convoy, the elderly were asked to recall the changes regarding the functional social support compared to their younger times.

3.1.2 Socioemotional Selectivity Theory

Socioemotional selectivity theory (SST) was first put forward by Carstensen and her colleagues in 1999. It claims that the perception of time plays a fundamental role in the selection and pursuit of social goals (Carstensen et al., 1999). In light of this life-span theory, when people regard the future as open-ended, they tend to pursue on the acquisition of knowledge, achieve their goals that aimed at personal development, and establish new social contacts that could be helpful in the future (Löckenhoff & Carstensen, 2004). Generally speaking, people in their youth time incline to be affected by this notion. When individuals are stepping into their later life, their thinking may have changed. When the future is limited, the perception of time would drive the old to focus more on their emotional demands. This includes goals that are concentrated on regulating one's emotions by avoiding negative states, intensifying positive states and making themselves more adaptable to the changing environments (Löckenhoff & Carstensen, 2004). This kind of intrapersonal inclination is more present-oriented; however, for individuals in the younger stage, their goals are probably more future-oriented.

Regarding the specific study, this theory postulates that people in old age may seek for more close social partners who can provide emotional support; on the contrary, the expansion of social network including peripheral social partners may be given less priority (Fung, Stoeber, Yeung, & Lang, 2008). The implications may emphasize that for the older people, more communication and interaction with close relationships would be appropriate, which might have a positive effect on the psychological adaptation to the aging process. As for the potential inspiration to the current research, this theory is assumed to be capable to generalize a series of research hypotheses in terms of the structural and functional support properties in old age. The implications of SST may shed light on the demands for emotional support and companionship support in later life. Additionally, another pivotal point from the implication of SST may be derived from the coping process of the elderly. Coping could help the elderly adjust to the aging process, which is considered to play an imperative role in older people's psychological adaptation. In line with the SST, it could be predicted that the elderly might tackle with their emotional demands selectively, prioritize personally relevant and positive affairs and try to get rid of potential negative feedbacks (Löckenhoff & Carstensen, 2004).

In a word, SST provides an explicit and direct access to examine elder individuals' social networks, which is highly associated with the research contents of my study.

Specifically, the theory elucidates the function of close and peripheral social partners in late life, which lends theoretical evidence to the proposition of research hypotheses.

3.1.3 Hierarchical Compensatory Model

In social gerontology, the role of the family has been well documented as people have lived longer, especially in the context that the older people are in need of care. To help explain individual differences in support networks and what happens when certain key members from these networks are missing (e.g., family members), Cantor (1979) proposed the hierarchical compensatory model. According to Cantor (1979, 1991), older adults have a clear preference of helping sources when they need assistance. Specifically, this conceptual scheme suggests that regardless of the type of assistance, the spouse is the preferred helping source. If a spouse is not available, then the elders might turn to their children. Similarly, if neither a spouse nor a child is present, then older adults elect to go to relatives and friends for aid (Krause & Jay, 1991).

The order of preference in the choice of one's source of support (e.g., kin, then non-kin, and then formal organizations) is closely related to the idea of compensation in social relationships (Antonucci, Fiori, Birditt, & Jackey, 2010). It addressed the role of peripheral social partners when kin is not available to provide social support. In light of the implication of the hierarchical compensatory model, the present study attempts to explore the support provided by both family members (the innermost circle) and peer groups (the outmost circle) and to examine whether the compensatory role of peer groups is significant.

3.1.4 Strengths Perspective

In social work practice, the strengths perspective is widely adopted (Saleebey, 1996). Strengths-based approaches value the capacity, skills, knowledge, connections and potential in individuals and communities. Applied in the present study, aging is not regarded as a process or stage on which problems, illness and losses converge. The strengths perspective helps to interpret the transition in aging from theories emphasizing the inevitable decline of human capacity to a series of notions that address a positive aging process with empowerment and resilience (Langer, 2004).

Inspired by the strengths perspective, older adults are encouraged to be actively involved in the community and build on family and community interactions and knowledge (Gilchrist, 2009). Additionally, researchers have found that by encouraging pride in achievements and a realization of what people have to contribute, the elderly in communities could generate increased confidence in their ability to be producers instead of recipients of development (Foot & Hopkins, 2010). Based on the implications derived from strengths perspective, the present study could be guided by articulating or building up the elders' capacities or assets. In regard to the components of positive aging, the elderly are recommended to mobilize the resources in the circumstances to maintain an active role in the community. Additionally, the older adults are expected to recognize their specific capacities with the help of strengths-based social work practitioners. With the strengths perspective, older adults may believe that they are capable of coping with stressors, troubles and losses, leading to resilient adaptation in later life.

3.2 Conceptual Framework

This section explicates the conceptual framework of the main research variables in accordance with the previous statement and analysis of relevant literature. The framework of social support network and the construct of positive aging are presented respectively. Afterwards the integrated conceptual framework of this present study is illustrated.

3.2.1 Conceptual Framework of Social Support Network

As is explicitly described in Chap. 2, social support network is proposed as the main independent variable in my study. Specifically, the operational definition of social support network may be elaborated as the combination of the portions which could provide support to the person, containing structural components and functional components. The structural framework of support network to be studied in the current research is drafted as follows (Fig. 3.1).

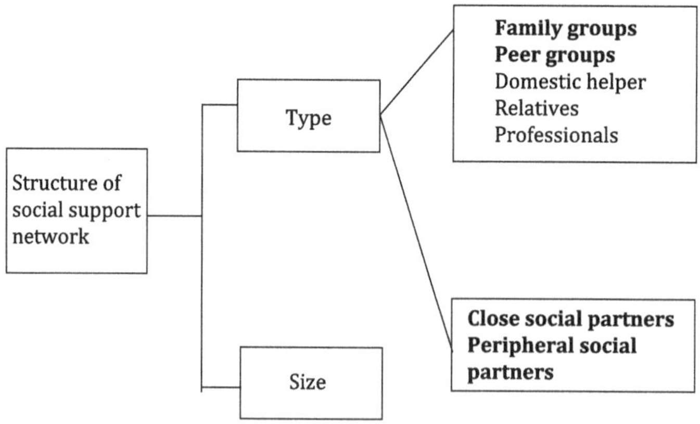

Fig. 3.1 Conceptual framework of structural components of social support network

As the above figure depicts, the structure of social support network has two dimensions—type and size. To draw a more detailed classification, support network types are considered to have two categories. According to the classification of the nature of support providers, five types of social networks could be generalized. Family groups in this study usually refer to the primary family members, including spouse, adult children, grandchildren, and children-in-law. Peer groups in this study are operationalized as a series of individuals considered equal exchanging assistance and support with each other, including neighbors, colleagues, former classmates, friends and community center members, etc. (Dennis, 2003). Among the social relation types, friendships often provide emotional support and companionship in later life because the interaction process may embed more norms of reciprocity and resources (Lang et al., 2009). In the present study, there is no distinct difference between the term *friends* and *peers*. The term *peer* is adopted in the present study because *peer* could best represent the "equal" interpersonal relations with others.

Domestic helper is defined as the full-time or part-time, local or non-local helper, which is employed to serve for the elderly or the household. Relatives are defined as the kin except for the primary family members. Professionals are operationalized as the hospital or clinic doctors, social workers, nursing workers, social organizations (e.g., community center, day care center, etc.).

In the present study, the researcher will focus on two types of social networks, namely, family groups and peer groups.

Size of social support network is operationalized as the number of members under each specific type of network. The social network size stands for the total number of one's social support networks nominated by the respondents. In line with the convoy model of social support and SST, the size of close social network is operationalized as the number of support providers who are regarded as vital or very important to the respondents, which are placed in the inner circle of the convoy. The size of peripheral social network is operationalized the social relations who are not very important but still non-ignorable, which pertain to the middle and the outermost circles.

In terms of the functional components of social support network, the present study aimed to examine four functions, namely, emotional support, companionship support, instrumental support and informational support. Emotional support involves providing love, empathy, caring, listening and trust, as well as those providing support which could release the negative mood of respondents. Companionship support describes a kind of support that could satisfy respondents' inner desire to be accompanied by others, from which to acquire the belongingness. Instrumental support refers to kinds of services, behaviors and other non-material forms of support to directly help the person in need, such as caregiving, curing diseases, etc. Informational support is defined as providing useful information which could be applied to solve problems, acting as a medium to provide assistance and guidance to persons, such as knowledge, skills, advices and education, etc.

In terms of the concept of emotional closeness, the present study proposes to examine the family emotional closeness and peer emotional closeness.

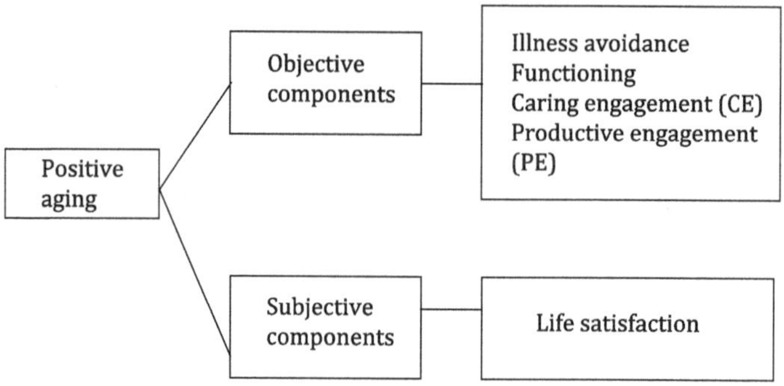

Fig. 3.2 Conceptual framework of positive aging

3.2.2 Conceptual Framework of Positive Aging

The conceptual framework illustrates positive aging, which is composed of five components. In the illustration of positive aging, the conceptual framework of this concept has five components. As documented in previous chapters, the present study aims to adopt the two-factor model of successful aging (Pruchno et al., 2010) and the extended construct of successful aging adjusted to Chinese older adults (Ng et al., 2011). The conceptual framework is presented as follows (Fig. 3.2).

As is outlined, the objective components of positive aging include four dimensions, namely, illness avoidance, functioning, CE and PE. Illness avoidance is defined as low probability of disease and disease-related disability. Functioning refers to high cognitive and physical functional capacity, the potential and competence an old person could achieve in later activities. CE is operationalized as showing love and care to primary family members and relatives, and assisting in household maintenance. PE is operationalized as undertaking paid or non-paid work and activities that can produce social or economic values.

3.2.3 Integrated Conceptual Framework of the Study

As Fig. 3.3 shows, the present study adopts positive aging as the dependent variable. Social support network and coping are independent variables. Additionally, coping is hypothesized to serve as a mediator in the relationship between social support network and positive aging. Furthermore, a couple of demographic factors (e.g., age, education, gender, perceived SES, employment of DH) are selected as covariates.

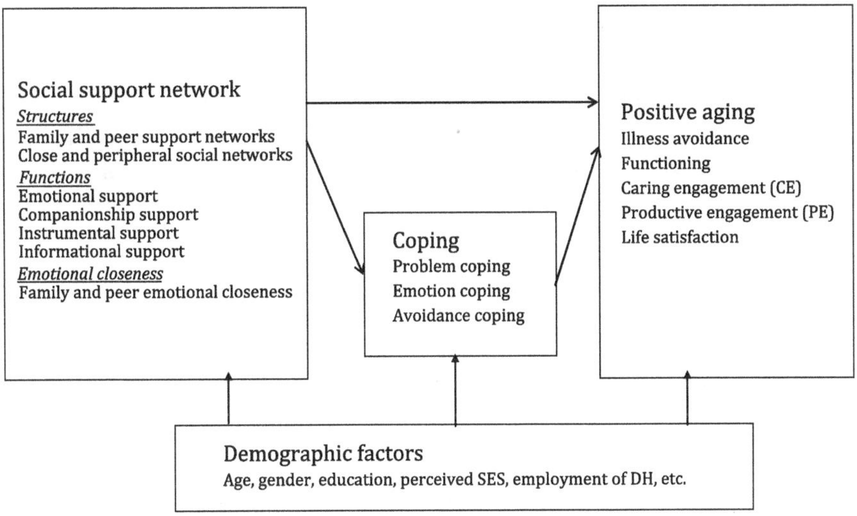

Fig. 3.3 Integrated conceptual framework of the study

3.3 Hypotheses

From the empirical studies and theories examined in previous sections, three sets of hypotheses are postulated. The first set is on the social support network, age, and positive aging inspired by the convoy model of social support and SST. The second set is on coping strategies and its mediating role in the relationship between social support network and overall positive aging. The third set is on the demographic factors.

3.3.1 Hypotheses Set 1

The convoy model of social support describes the structural and functional components of social relations. The present study aims to examine the structures and functions of convoy model and raises research questions and hypotheses in light of the theoretical implications of convoy model. Empirical studies of the convoy model have offered insight into the prevalent roles of family and peers and their contribution of well-being in later life. A set of hypotheses are proposed in terms of the structural and functional properties of family and peer networks.

Hypothesis 1.1 The size of family network has a positive effect on positive aging.
Hypothesis 1.2 The size of peer network has a positive effect on positive aging.
Hypothesis 1.3 The social network size has a positive effect on positive aging.

Hypothesis 1.4 Functional family support has a positive effect on positive aging.
Hypothesis 1.5 Functional peer support has a positive effect on positive aging.

SST explains the effect of age differences on individual's social network composition. As stated in previous chapters, SST argues that when a person steps into later life, one's social network would be narrowed due to their increasing demands of emotional closeness and decreasing need of open-ended pursuit in future goals. Therefore, emotionally close social partners are preferred because they are more likely to provide predictable emotional support and companionship (Fung et al., 2008). Based on the implications of SST, a set of hypotheses is proposed.

Hypothesis 1.6 Age is correlated with size of close social network and peripheral social network, respectively.
Hypothesis 1.7 Age is negatively correlated with social network size.
Hypothesis 1.8 The size of peripheral social network is smaller than that of close social network.
Hypothesis 1.9 Older adults receive more emotional support than that of ten years ago.
Hypothesis 1.10 Older adults receive more companionship support than that of ten years ago.
In terms of the effect of emotional closeness on positive aging, the present study proposed a set of hypotheses that state the hypothetical relations between them.
Hypothesis 1.11 Family emotional closeness has a positive effect on overall positive aging.
Hypothesis 1.12 Peer emotional closeness has a positive effect on overall positive aging.
Hypothesis 1.13 The size of close social network has a positive effect on overall positive aging.

3.3.2 Hypotheses Set 2

In light of the evidence mentioned in previous work, a set of hypotheses on coping is proposed as follows. Hypotheses 2.1–2.3 are about the effect of age on three types of coping strategies respectively, inspired by Ben-Zur's (2002) findings. Hypotheses 2.4–2.6 are related to the contribution of three types of coping to overall positive aging. Hypotheses 2.7–2.12 presented the hypothetical mediating mechanisms between coping, social support network and positive aging.

Hypothesis 2.1 Age is positively correlated with emotion coping.
Hypothesis 2.2 Age is positively correlated with avoidance coping.
Hypothesis 2.3 Age is negatively correlated with problem coping.
Hypothesis 2.4 Problem coping has a positive effect on positive aging.
Hypothesis 2.5 Emotion coping has a positive effect on positive aging.

Hypothesis 2.6 Avoidance coping has a positive effect on positive aging.
Hypothesis 2.7 Problem coping can mediate the effect of the size/function of family support network on positive aging.
Hypothesis 2.8 Emotion coping can mediate the effect of the size/function of family support network on positive aging.
Hypothesis 2.9 Emotion coping can mediate the effect of the size/function of peer support network on positive aging.
Hypothesis 2.10 Problem coping can mediate the effect of family/peer emotional closeness on positive aging.
Hypothesis 2.11 Emotion coping can mediate the effect of family/peer emotional closeness on positive aging.
Hypothesis 2.12 Avoidance coping can mediate the effect of family/peer emotional closeness on positive aging.

3.3.3 Hypotheses Set 3

In this section, a set of hypotheses on the relationship between selected demographic factors, independent variables and dependent variable are raised. Hypotheses 3.1–3.2 are about the age effect on positive aging and the size of social network. Hypotheses 3.3–3.7 are related to the impact of education on coping strategies, the size of social network and positive aging, based on the argument that people who had a higher education level possessed more coping resources (Blanchard-Fields & Irion, 1988). Hypotheses 3.8–3.10 focused on the effect of perceived SES on positive aging, social support network, and the employment of DH.

Hypothesis 3.1 Age is negatively correlated with positive aging.
Hypothesis 3.2 Age is positively correlated with the employment of domestic helper.
Hypothesis 3.3 Education is positively correlated with Emotion coping.
Hypothesis 3.4 Education is positively correlated with Avoidance coping.
Hypothesis 3.5 Education is positively correlated with Problem coping.
Hypothesis 3.6 Education is positively correlated with positive aging.
Hypothesis 3.7 Education is positively correlated with social network size.
Hypothesis 3.8 Perceived SES is positively correlated with social network size.
Hypothesis 3.9 Perceived SES is positively correlated with positive aging.
Hypothesis 3.10 Perceived SES is positively correlated with employment of DH.

In this study, social support provided by DHs (e.g., full-time live-in FDHs, full-time and part-time local DHs) is one of the research questions. In line with the findings showing that support provided by DHs could facilitate community-based and home-based care services for the elderly, a group of hypotheses concerning the role of DH related to positive aging has been proposed.

Hypothesis 3.11 Employment of DH is positively correlated with positive aging.
Hypothesis 3.12 Among the older adults who employ DHs, satisfaction with domestic helper's care provision is positively correlated with positive aging.
Hypothesis 3.13 Older adults who employ DH have less intention to stay at residential care homes, compared with those without DH.

3.4 Summary

Drawing from the existing literature and theoretical basis in the area of social gerontology, the conceptual frameworks were postulated for the present study in this chapter. They included two theoretical frameworks on the construct of social support network and positive aging, and the integrated analytical framework on the main study. In light of the conceptual frameworks, forty-four hypotheses were then proposed.

First, the construct of the dependent variable—social support network was displayed. As a complex concept endowed with many attributes, the conceptual framework of social support network systematically presented the structural and functional properties that were about to be examined in this study. Among a variety of network types, family and peer networks were selected as the research focus in this study. Additionally, emotional closeness, which is rarely explored in past studies, was highlighted.

Second, the renewed conceptual framework of the dependent variable—positive aging was presented. It contained four objective components (e.g., illness avoidance, functioning, CE and PE) and one subjective component (e.g., life satisfaction).

Third, the integrated analytical framework which combined the independent variables, dependent variables, mediators and demographic factors was proposed. The integrated framework depicted each construct of main variables and the hypothetical relationships among variables.

Finally, based on the conceptual frameworks and reference from previous literature and theories, a total of 38 hypotheses were proposed in this chapter.

References

Antonucci, T. C., & Akiyama, H. (1987a). An examination of sex differences in social support in mid and late life. *Sex Roles, 17*(11), 737–749.

Antonucci, T. C., & Akiyama, H. (1987b). Social networks in adult life and a preliminary examination of the convoy model. *Journal of Gerontology, 42*(5), 519–527.

Antonucci, T. C., & Akiyama, H. (1995). Convoys of social relations: Family and friendships within a life span context. In R. Blieszner & V. H. Bedford (Eds.), *Handbook of ageing and the family* (pp. 355–371). Westport, CT: Praeger Publisers.

Antonucci, T. C., Fiori, K. L., Birditt, K., & Jackey, L. M. (2010). Convoys of social relations: Integrating life-span and life-course perspectives. In R. M. Lerner, M. E. Lamb, & A. M. Freund (Eds.), *The handbook of life-span development* (Vol. 2, pp. 434–473). Hoboken, NJ: Wiley.

Ben-Zur, H. (2002). Coping, affect and aging: The roles of mastery and self-esteem. *Personality and Individual Differences, 32*(2), 357–372.

Berges, I. M., Dallo, F., DiNuzzo, A., Lackan, N., & Weller, S. C. (2006). Social support: A cultural model. *Human Organization, 65*(4), 420–429.

Blanchard-Fields, F., & Irion, J. C. (1988). Coping strategies from the perspective of two developmental markers: Age and social reasoning. *The Journal of Genetic Psychology, 149*(2), 141–151.

Cantor, M. H. (1979). Neighbors and friends: An overlooked resource in the informal support system. *Research on Aging, 1*(4), 434–463.

Cantor, M. H. (1991). Family and community: Changing roles in an aging society. *The Gerontologist, 31*(3), 337–346.

Carstensen, L. L. (1992). Social and emotional patterns in adulthood: Support for socioemotional selectivity theory. *Psychology and Aging, 7*(3), 331–338.

Carstensen, L. L., Isaacowitz, D. M., & Charles, S. T. (1999). Taking time seriously: A theory of socioemotional selectivity. *American Psychologist, 54*(3), 165–181.

Cumming, M. E. (1964). New thoughts on the theory of disengagement. *New Thoughts on Old Age* (pp. 3–18). Berlin: Springer.

Dennis, C. L. (2003). Peer support within a health care context: A concept analysis. *International Journal of Nursing Studies, 40*(3), 321–332.

Depner, C. E., & Ingersoll-Dayton, B. (1988). Supportive relationships in later life. *Psychology and Aging, 3*(4), 348–357.

Erikson, E. H. (1950). *Childhood and society*. New York, NY: Norton.

Foot, J., & Hopkins, T. (2010). *A glass half-full: How an Asset approach can improve community health and well-being*. London: Improvement and Development Agency.

Fung, H. H., Stoeber, F. S., Yeung, D. Y. L., & Lang, F. R. (2008). Cultural specificity of socioemotional selectivity: Age differences in social network composition among Germans and Hong Kong Chinese. *The Journals of Gerontology, Series B: Psychological Sciences and Social Sciences, 63*(3), 156–164.

Gilchrist, A. (2009). *The well-connected community: A networking approach to community development*. Bristol: Policy Press.

Hill, R. D. (2005). *Positive aging: A guide for mental health professionals and consumers*. New York: W. W. Norton & Co Inc.

Kahn, R. L., & Antonucci, T. C. (1980). Convoys over the life course: Attachment, roles and social support. In P. B. Baltes & O. Brim (Eds.), *Life-Span development and behavior* (Vol. 4, pp. 253–286). New York: Academic Press.

Krause, N., & Jay, G. (1991). Stress, social support, and negative interaction in later life. *Research on Aging, 13*(3), 333–363.

Lang, F. R., Wagner, J., & Neyer, F. J. (2009). Interpersonal functioning across the lifespan: Two principles of relationship regulation. *Advances in Life Course Research, 14*, 40–51.

Langer, N. (2004). Resiliency and spirituality: Foundations of strengths perspective counseling with the elderly. *Educational Gerontology, 30*(7), 611–617.

Levitt, M. J., Weber, R. A., & Guacci, N. (1993). Convoys of social support: An intergenerational analysis. *Psychology and Aging, 8*(3), 323–326.

Löckenhoff, C. E., & Carstensen, L. L. (2004). Socioemotional selectivity theory, aging, and health: The increasingly delicate balance between regulating emotions and making tough choices. *Journal of Personality, 72*(6), 1395–1424.

Lynott, R. J., & Lynott, P. P. (1996). Tracing the course of theoretical development in the sociology of aging. *The Gerontologist, 36*(6), 749–760.

Ng, S. H., Cheung, C. K. J., Chong, A. M. L., Woo, J., Kwan, A. Y. H., & Lai, S. F. Y. (2011). Aging well socially through engagement with life: Adapting Rowe and Kahn's model of successful aging to Chinese cultural context. *The International Journal of Aging and Human Development, 73*(4), 313–330.

Pruchno, R. A., Wilson-Genderson, M., & Cartwright, F. (2010). A two-factor model of successful aging. *The Journals of Gerontology Series B, 65*(6), 671–679.

Saleebey, D. (1996). The strengths perspective in social work practice: Extensions and cautions. *Social Work, 41*(3), 296–305.

Tornstam, L. (2000). Transcendence in later life. *Generations, 23*(4), 10–14.

Chapter 4
Method

In this chapter, the methodology of the mixed-method study (i.e. in-depth interview and survey) is presented. This chapter has four parts. Part one introduces the research design of the mixed-method study. Part two presents the research design of the pilot study, followed by the research procedures of the main study in part three. Then the process of scale construction and validation for the main study is presented.

4.1 Research Design: A Mixed-Method Study

The present study aims to examine the properties of social support network, coping and positive aging among the community-dwelling elders in Hong Kong. To acquire a reliable and comprehensive result, a mixed method study was adopted in this study, which was composed of a pilot study with a main study.

Tashakkori and Teddlie (2003) suggested a definition of mixed methods study as "…involves the collection or analysis of both quantitative and qualitative data in a single study in which the data are collected concurrently or sequentially, are given a priority, and involve the integration of the data at one or more stages in the process of research" (p. 212).

A mixed methods study usually encompasses a combination of qualitative and quantitative methods which could be conducted either concurrently or sequentially in terms of the need of research purpose. It is generally supposed that combining quantitative or qualitative approaches in one study, or in a series of subsequent research studies, is more useful than a separate application in social science research. A mixed methods research could mobilize the advantages of both kinds of methods in a study or work as complimentary approaches to explore research topics. As discussed, the strengths of quantitative methods are that they can produce factual, reliable outcome data that are usually used to generalize to potential larger population. However, the strengths of qualitative methods may include that they can generate rich,

S. Liu, *Social Support Networks, Coping and Positive Aging Among the Community-Dwelling Elderly in Hong Kong*,
https://doi.org/10.1007/978-981-13-3783-3_4

detailed, valid process data or provide contextual understanding of health behavior and program results (Steckler, McLeroy, Goodman, Bird, & McCormick, 1992).

As Steckler et al. (1992) put, a typical and frequently-used mixed methods model is to adopt qualitative methods to develop quantitative measures in a study. This kind of mixed methods study is not conducted simultaneously, but sequentially as proposed. This sequential design is often accompanied with exploratory and explanatory objectives, usually showing the priority on quantitative study or both. Qualitative study is conducted first to acquire first-hand data and then to provide guidelines for quantitative study measures. After the quantitative study in the second phase is completed, data will be collected and analyzed together with the qualitative results to interpret the whole findings (Tashakkori & Teddlie, 2003). There are a couple of specific reasons needed to throw light upon why a mixed-method is chosen for this study. First, in general, a mixed-method is popular in social science research nowadays since it endows a more holistic view to examine the research topic, to some extent, which diminishes the potential bias in actual study. The survey study could explicitly unravel the correlational relationships between the variables via statistics analysis and model construction; while a qualitative study could capture the human behaviors and attitudes through a direct interaction between the researcher and respondents. In the present study, the target informants were Hong Kong older Chinese. One of the research objectives was to identify the specific facets related to aging issues among Hong Kong older adults. Thus, a survey study was a good choice which systematically gathered quantitative data from a relatively large number of subjects and allowed kinds of statistical analysis due to the standardized answers from informants. Additionally, in consideration of the features of respondents in this study, many of them might age 75 or above, namely the old-old, could not be self-capable in filling the questionnaires. Second, one ethnic question should be noticed in relation to the researcher. The target research site Hong Kong, is a special administrative region of the People's Republic of China, which possesses a different social system and cultural endorsement from regions in mainland China. Thus, to conduct a study in Hong Kong for the researcher—a mainland Chinese might yield several problems that need to throw light upon. Before the study was carried out, the researcher did not have adequate knowledge of the research questions of the study. Thereby, it was impossible to collect scientific research findings using the questionnaire which was solely constructed based on the subject perception of the researcher and the previous literature. Therefore, the implementation of the pilot study could collect qualitative data related to the research questions, and provide preliminary information to interpret the research topic. What's more, the findings from the qualitative study could also be utilized to facilitate questionnaire construction (e.g., add or reduce items, find more appropriate presentation of questions). By this way, the qualitative study could pave way for the following survey and provide guidelines in scale revision.

By and large, the adoption of mixed methods could get the research to approach the proposed research questions in different perspectives, and to enable researchers to cross-check the results generated by different methods to improve the persuasiveness of research findings (Peace & British Society of Gerontology, 1990). Thus, the flow of overall research design was illustrated as below (see Fig. 4.1).

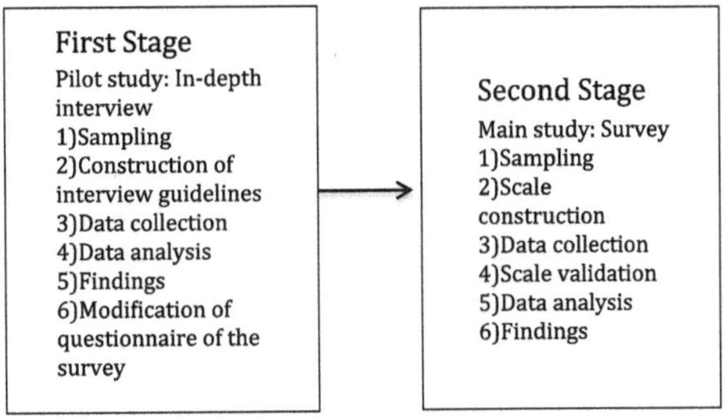

Fig. 4.1 Flow of research design of the mixed-method study

4.2 Pilot Study: In-Depth Interview

4.2.1 Sampling

Unlike the sampling method in quantitative research using careful and mathematical calculation and design, the qualitative study may show some distinct differences. Compared to quantitative sampling, the qualitative sampling is more flexible with the concentration of information rich and in-depth (Li, 2010).

Taking account of the specific situation of the study, the in-depth interview could not be conducted without any theoretical foundations. Moreover, a preliminary conceptual framework consisting of selected important variables have already been proposed after a comprehensive and systematic literature review on the research topic. Consequently, in the first stage, the researcher constructed the interview guidelines on the basis of the proposed framework and endeavored to cover the significant and puzzling questions which remained further identification. Among the methodologies in qualitative studies (e.g., participant observation, content analysis, life-story book), the in-depth interview was more purposeful, which was more appropriate to be carried out in this study, considering the availability of resources and time.

With regard to the sampling size of the in-depth interview, it was not as accurate as a survey to determine how many informants should be involved. Generally, there was not a consistent and standardized method to calculate the sampling size. Nevertheless, it was certain that a qualitative study could almost not adopt a probability sampling whose aim was to achieve representativeness in a larger population (Teddlie & Yu, 2007). Therefore, it was more accessible to employ a non-probability sampling method. According to the experiences in past studies, purposive sampling and convenience sampling were most frequently used in qualitative studies. Considering that a few biases may exist in a convenience sample (e.g., the uncertainty of informants' heterogeneity), a purposive sampling method was a better approach for the in-depth interview.

In the process of purposive sampling, the researcher was able to actively select qualified respondents the research question. The purposive sampling can involve the requisite variables in interview guidelines which were based on the researcher's practical knowledge of the research area, the available literature and evidence from the study itself (Marshall, 1996).

In the first stage, the researcher selected two DECC in Wan Chai District and Sham Shui Po District of Hong Kong respectively. There were two reasons for the selection method. First, the two selected districts had shown obvious differences in terms of *Social Economic Status*. In this study, the status of *Median monthly household income* was chosen as the indicator to judge. According to the *Latest Population and Household Characteristics* (Census and Statistics Department, 2012), the median monthly household income was $17,500 in 2010, and dramatic diversities existed among the 18 administrative districts in Hong Kong. Of all the districts, Wan Chai District ranked the first place in median monthly household income with $28,000, and Sham Shui Po ranked the last in that indicator with $13,700. Second, in addition to the huge difference in monthly household income, both districts embraced a high percentage of older population. In 2006, people aged 65 or above accounted for 12.4% of the total population in Hong Kong, moreover, Wan Chai District and Sham Shui Po District had a rate of 14.1 and 16.7% respectively (Census and Statistics Department, 2007). Therefore, both the districts could indicate their representativeness with a high density of older adults in Hong Kong.

4.2.2 Sample Size Determination

An appropriate sample size for a qualitative study is one that adequately answers the research questions (Marshall, 1996). In practice, the number of required subjects usually becomes obvious as the study progresses, as new categories, themes or explanations stop emerging from the data, which is also a distinct requirement of grounded theory. However, in regard to my study, the main aim was not to explore a new social phenomenon or induce a specific theory or model from the findings, but an explorative extension of existing knowledge and provision with advantageous modification for quantitative measures in the next stage. Thus, the sampling size was not necessarily that large.

Additionally, the response rate could be controlled since the interviewees were affirmed in advance and the interviewers could be equipped with necessary skills to conduct in-depth interviews. Hence, it was not necessary to strictly calculate the failure rate and incomplete or illegible responses, which, to a large extent, diminished the scale of sampling frame.

For the qualitative method—in-depth interview, the sampling size could not be too large because it yielded very rich data, and even with small numbers, the researcher can be left with several hundred pages of interview texts to analyze. In addition, taking consideration of the restriction of resources (e.g., research grant, the number of interviewers and the tight schedule) the researcher was unable to conduct a relatively large scale of interviews.

Table 4.1 Sampling criteria and dimensions of variables of the pilot study

Criteria	Dimensions	
Age	60–74	75 and above
Health condition	Self-care capable	Self-care incapable
Gender	Female	Male
District	Wan Chai District	Sham Shui Po District

Considering the objective of informant selection, it was expected to get a sample representing different characteristics of Hong Kong older adults associated with the research objectives in this study. There were several facets needed to take into account of when choosing a sample. Firstly, the young-old and the old-old were included to expand the heterogeneity. Secondly, older people in different health conditions (e.g., in good health, frail or home-bound) were selected. Thirdly, the sample should have a balanced proportion of gender.

Based on the above disciplines, district, age, health condition and gender constituted the selection criteria of the informants. According to the dimensions under each factor, each factor was divided into two standards during the selection process. Table 4.1 illustrates the details of respondents in the whole sample in terms of the dimensions under selection criteria.

Thus, the total sampling size for the in-depth interview could be calculated as $2 \times 2 \times 2 \times 2 = 16$.

4.2.3 Informants Recruitment

To facilitate the in-depth interview, the recruitment of informants was conducted subject to the below criteria. First, all the interviewees showed consent to participate in the interview. A consent form explaining the objectives of the study, confidentiality, reference persons and procedures had been sent to them in advance. Then the social workers working in DECC introduced the research purpose to them in advance to obtain trust. Second, all the selected respondents have demonstrated basic capacities to communicate either in Cantonese or Mandarin verbally.

4.2.4 Construction of Interview Guidelines

By virtue of the preliminary conceptual framework containing properties of social support network, positive aging and coping strategies, the interview questions were divided into three parts. In the first part, respondents were mainly asked to talk about social support network concerning the role of family members, friends and domestic helpers (if any) in support provision. Furthermore, questions were also set to explore the needs and attitudes towards emotional, instrumental, companionship and informational support. Additionally, informants were invited to recall the support network changes in comparison with the younger times. In the second part, coping was enquired to attain information about the types of coping strategies they would use when confronted with stressors brought by aging. Lastly, in the third part, their perceptions or attitudes towards the components of positive aging were discussed.

4.2.5 Data Collection

Before the recruitment of informants, the research had contacted the executives and social workers of the community centers. Upon receiving approvals, the researcher sought help from the social workers to recruit interviewees. Interviews were conducted in the case counseling room of the center or respondents' home.

The interviews were recorded after the researcher got the permission from respondents. After an interview, data were transcribed into Chinese texts immediately after it was finished so as to start data analysis sooner. During the process of interview, memos were written immediately to record the perceptions of the researcher and observations from the conversation.

4.2.6 Data Analysis

Generally, there are several approaches of data analysis to qualitative inquiry. Grounded theory is a systematic methodology which extracts important points (namely codes) from the texts of observational data (Babbie, 2010). It needs several rounds of systematic coding until the themes and categories discovered from the data could construct theories. Specifically, there is no hypothesis or theoretical framework before the study conducted in grounded theory. Another method of data analysis in qualitative research is thematic analysis, which is similar with grounded theory. Thematic analysis tries to analyze the themes which capture something important about the data in relation to the research questions and represent patterned responses or meanings within the data set (Braun & Clarke, 2006). They both attempt to represent a view of reality via systematically coding and then group the effective elements together and realize the reconstruction of theories. Different from grounded theory, thematic analysis is conducted with the existence of proactive themes or patterns, and then conclusions are drawn from the combined and catalogued data. It was more appropriate to adopt thematic analysis for the purpose of obtaining first-hand information regarding the research topic.

4.2.7 A Review of the Qualitative Study

A review of the qualitative study was presented to acknowledge the critical issues that were embedded in the research design, data collection and data analysis. In light of the Guidelines for Critical Review for Qualitative Studies (Letts et al. 2007), the following section described a review of the research purpose, study design, sampling, ethnic issues and data analysis.

The purpose of the qualitative study was to acquire first-hand information in relation to the research focus of the study. The research question of the qualitative study was to get an overall image of the properties of social support network (e.g., types and functions), positive aging and coping. The qualitative study chose an in-depth interview design. A number of issues were considered in determining the appropriateness of the design. First, in-depth interview was an effective and straightforward approach to meet the research purpose of the study. The research could collect verbal data directly from the conversation with interviewees. Second, the research was able to freely select interviewees according to some criteria so that voices from diverse informants could be collected. Third, interviews can be done relatively quickly, with little expense, and are useful when a particular issue needs to be explored in depth.

With regard to sampling, the qualitative study used a purposive sampling method which was implemented in accordance with specific criteria. The objective was not to acquire a representative sample, hence, criteria was only proposed to obtain a relatively diverse sample. In addition, the ethical issues of the informants group were also considered. The informants were DECC members and most of the interview were conducted in DECC. All the participants were selected on a voluntary basis and fully acknowledge the interview procedure and rules. The informants were aware that they could refuse to answer the questions that were raised by the researcher. Each of them sighed the consent form and was presented a small gift as incentive. The interview was carried out under the systematic arrangement of the social worker working in DECC. The interview did not disturb the regular activities of DECC. As to the two interviews which were conducted at home, the two informants had approved the home visit upon knowing about the purpose of the research before the interview started. In the process, the researcher has showed respect to all the informants' participation. In the meantime, the research was aware that the data collected from the participants were only reflected from the lenses of the selected DECC members. The findings could not be generalized to all the DECC members.

As to data analysis, the transcription, translation and thematic analysis were done by the researcher, thus leading to some research bias. Although the researcher was able to speak Cantonese, which was the communication language, some inappropriate interpretation might occur because of the cultural diversity or missing of conversation contents. Additionally, the Chinese transcript was translated to English all by the researcher, leading to probable inaccurate translation. Furthermore, a few data might not be classified into the selected themes because of the subjective judgment of the researcher.

This review of the qualitative study helps to understand the research findings in a more objective way. Then the findings were further interpreted and extracted for the questionnaire design of the survey.

4.3 Main Study: Cross-Sectional Survey

4.3.1 Sampling

As to a survey research, the sampling method is very important as a technique and process of selecting an appropriate sample, or a representative branch of a population for the purpose of estimating characteristics of the whole population (Babbie, 2007). Additionally, the quality of the selected sample is crucial in terms of its composition and representativeness. Generally, on the condition that all other things being equal, the larger the sample size, the more reliably a relationship can be identified. However, the selection of sampling method and the final determination of sample size must be affected by a series of factors in each specific survey, such as the time and grant needed to administer a survey. Considering the specific situation for the researcher to conduct the present survey, it was proposed that a multi-stage sampling method was employed and a convenient sample of 400 was targeted.

4.3.2 Sample Size Determination

Considering the target population was a large amount over 940,000 (Population Census, 2012a), it was set that we choose a margin of error (ME) less than 5%, that is, the confidence level was 95%. And typically we assumed the confidence interval as 5. For the prior judgment of the correct value p, at this stage since we had no initial estimate of p, we adopted the convention that a conservation assumption $p = 0.5$ was allowed for being unknown a priori. Though the target population was huge, it should be noticed that the study was not able to obtain a large sample because of the restriction of energy and resources. Therefore, the study adopted Cochran's sample size formula for categorical data (Cochran, 1977). In line with the formula of sample size calculation, the proposed number of sample units was 384.

$$\mathrm{ME} = Z\sqrt{\frac{P(1-P)}{N}}$$

In the above formula,

(I) ME stands for the desired margin of error.
(II) Z represents the Z-score, which is the parameter matching the confidence interval.

(III) P is the prior judgment of the correct value.
(IV) N is the to-be-found sample size.

As to the present study, ME equaled to 0.05, $Z = 1.96$ when the confidence interval was chosen as 5. Here p was assumed as 0.5. Over the calculation, finally we got N equals 384. To facilitate the operation in recruiting participants, a convenience sample of 400 was proposed. Finally, a total of 393 questionnaires were obtained.

Another issue that should be taken into account of is the statistical power for testing the hypotheses of the study. Power analysis allows us to make sure that we have looked hard enough to find it, or whether we could have enough power to reject the null hypothesis (Cohen, 1989). To calculate the statistical power of the sample size in testing hypotheses, a Power Analysis and Sample Size Software version 13 (PASS) was employed. In the survey, descriptive statistics and inferential statistics were carried out. Mediation analysis and hierarchical regression analysis were selected to test the hypotheses. Based on the results of hypotheses testing, the PASS was used to calculate the minimum sample size that was required to reach a satisfactory power.

In carrying out the calculation of power analysis, 0.8 was entered as the baseline of power and 0.05 for Alpha. Additionally, a number of four demographic variables (e.g., age, gender, perceived SES and educational background) were controlled in each step of hierarchical regression analysis, and there were a number of three independent variables at most in each step. Furthermore, the variance (R^2) that was explained by the regression model ranged from 0.03 to 0.1 approximately. Hence, the PASS finally calculated the minimum sample size to achieve the above requirements for hierarchical regression. Consequently, the power analysis revealed that a sample size of 357 achieved 80% power to detect an R^2 of 0.03 attributed to independent variable(s) using an F-Test with a significance level (alpha) of 0.05. Therefore, it could be inferred that the current sample size (393) was capable to meet the power in testing the hypotheses.

4.3.3 *Multi-stage Sampling Procedure*

Generally, it is very difficult to use random sampling method in Hong Kong due to some reasons such as the problem to get a full list of citizens. As for Hong Kong, it consists of 18 District Council districts, which are scattered in Hong Kong Island, Kowloon and New Territories. Considering the availability of resource and energy, a multi-stage sampling method was adopted.

In this survey, a three-stage sampling method was implemented. It should be noted that the three-stage sampling method could not generate a probability sample. Thus, a randomization was only used in the first stage to select districts. In the following stages, a convenient sampling was employed. In the first stage, a total of 18 districts were taken as the primary units for sampling. All the target population cases in each district were listed, together with the accumulative cases (see Table 4.2). To obtain a

Table 4.2 Distribution of population aged 65 or above in Hong Kong

Name of Districts	Population of older adults aged 65 or above	Accumulative No.
Central and Western	35,026	35,026
Wan Chai	23,834	58,860
Eastern	91,882	150,742
Southern	38,869	189,611
Yau Tsim Mong	44,690	234,301
Sham Shui Po	64,849	299,150
Kowloon City	60,415	359,565
Wong Tai Sin	74,036	433,601
Kwun Tong	101,463	535,064
Kwai Tsing	74,927	609,991
Tsuen Wan	39,392	649,383
Tuen Mun	47,220	696,603
Yuen Long	54,783	751,386
North	32,576	783,962
Tai Po	31,860	815,882
Sha Tin	72,285	888,107
Sai Kung	39,342	927,449
Islands	13,756	941,205
Total	941,205	941,205

Source Population Census (2012b). *Population by District Council District, Sex, Age Group and Educational Attainment (Highest Level Completed), 2011(B209)*

sample which could represent the characteristics of Hong Kong to a large extent, one third of the districts, that is, six districts were selected after balancing the restriction of finance and energy. Six random figures were chosen from Random Integer Generator. Then six districts were randomly determined, they were Sha Tin, Kwai Tsing, Sham Shui Po, Wai Chai, Kowloon City (repeated for the second time). In the second stage, all the DECCs in each of the selected districts were listed as the secondary sampling unit. After coordinating with the DECCs, invitation letter stating the survey purpose and procedures was mailed to each and every DECC of the six selected districts. The DECC which first approved the study request in each district was chosen as the site of study. Consequently, the six DECCs were Ma On Shan DECC (Sha Tin), SAGE Chan Tseng Hsi DECC (Kwai Tsing), Ho Kin DECC (Sham Shui Po), St James Settlement DECC (Wan Chai), Holy Carpenter Church DECC (Kowloon City), and Lok Man Alice Kwok Integrated Service Center (Kowloon City). In the third stage, invitation letters were distributed to all center members of the chosen DECC and the first 68 members who gave consent to participate in this survey were invited to the interview at the center on an appointed day.

4.3.4 Data Collection

Data collection started from November 2012 and ended in May 2013. A team of 35 undergraduates and postgraduates from social science disciplines (e.g., social work, psychology, sociology and criminology) from two local universities (e.g., City University of Hong Kong and The Hong Kong Polytechnic University) were recruited and trained as interviewers. Before conducting interviews, a 3-h training lesson was prepared to equip the students with systematic knowledge and skills of interviewing older people. Student interviewers rehearsed in pairs by taking turns to act as the interviewer and interviewee. Taking into account of the literal capacity of older adults, interviewers were required to read out the questions and to explain the guidelines to the respondents. Each interview lasted approximately 30–40 min.

4.3.5 Data Analysis

Statistical analyses were performed using IBM Statistical Package for the Social Sciences (SPSS) version 21.0 in conjunction with SPSS AMOS 20.0. Correlation analyses were carried out with the demographic factors and selected variables: age, gender, educational attainment, perceived SES, social support network, coping, and positive aging in order to provide an overview of the relations. Confirmatory factor analysis (CFA) for expanded PAI and Simplified Coping Style Questionnaire (SCSQ) were conducted to assess structural validity. Next, based on the preliminary findings from correlation analysis, hierarchical regression analysis using SPSS and path analysis using AMOS were carried out to examine the relationships between the above. Path analysis has the benefit that it can confirm the extent to which factors are mediated and provides estimates of the relative importance of direct and indirect factors. Thus, mediation models were constructed based on the previous hypotheses and on significant correlations using path analysis.

4.4 Scale Construction and Validation

4.4.1 Scale Construction

In line with the findings of the in-depth interview and existing literature on the measurement of main variables, the questionnaire for the survey was constructed. In this section, each part of the questionnaire was elaborated, including the selection of scales and modification of scales.

4.4.1.1 Screening Test and Demographic Variables

Eligible participants were required to age 65 years or above and to have sufficient cognitive ability to understand the questionnaire. Thus, Short Portable Mental Status Questionnaire (SPMSQ) was used at the beginning of the interview to screen for eligibility in cognitive functioning. SPMSQ is a 10-item test measuring the presence and the degree of intellectual impairment of the elderly with a dichotomous score of 0 and 1. Participants who get the right answer for one question will get one point. Pfeiffer (1975) suggested that participants with a score of 3–4 points of SPMSQ were diagnosed as suffering moderate intellectual impairment. In this study, *4* was adopted as the cutting point to differentiate the eligible participants.

Participants reported their age, gender (1 = male, 2 = female), marital status (1 = single, 2 = separate, 3 = married, 4 = divorced, 5 = widowed), educational attainment (1 = illiterate to 6 = higher education), perceived SES (1 = lower to 5 = upper). Additionally, participants were asked about the experience of employing domestic helper to undertake caring tasks including caregiving or household maintenance in the past year (0 = never employed, 1 = ever employed part-time or full-time domestic helper) and the types of services provided by DHs. Furthermore, they were asked to rate the satisfaction of services provided by DHs (1 = extremely not satisfied, 5 = extremely satisfied). Additionally, participants were asked whether they were willing to move out to residential care homes if their health deteriorates (1 = yes, 2 = not clear, 3 = no).

4.4.1.2 Structures of Social Support Network

In social gerontology, a couple of measurement tools were designed to scrutinize the structural properties of the individual's social support network. The Lubben Social Network Scale (LSNS) is a widely used measurement instrument to gauge social isolation in older adults by testing their perceived social support mainly from family members and friends. The original scale was designed by Lubben (1988). It consisted of an equally weighted sum of 10 items used to measure size, closeness and frequency of contacts of a respondent's social network (Description of the LSNS, 2011). Later on, Lubben and his colleagues delineated the revised version (LSNS-R), along with an abbreviated version (LSNS-6) and an expanded version (LSNS-18) for different professional research purposes respectively. In Hong Kong, the original version of LSNS has been adopted in empirical studies of the elderly in past studies. According to the exploration of the reliability and validity of LSNS, a study conducted by Chi and Boey (1992) indicated that the Cronbach's Alpha was 0.75.

As to the present study, it was proposed to employ LSNS-R as the measurement of support network. LSNS-R has a good reliability as measured by the internal consistency with a Cronbach's Alpha value of 0.78. The correlation coefficient between the original LSNS and LSNS-R was 0.68.

In this study, the convoy model was proposed to underpin the theoretical base of the structural properties of social support. Therefore, it was suggested that a specific instrument aiming to explore the convoy membership changes with age was needed. In accordance with Kahn and Antonucci (1980), the Social Convoy Questionnaire (SCQ) was utilized in this study to calculate the network size and composition. Participants were asked to nominate social partners who were important to them in support provision into three circles respectively. The inner circle indicated emotionally close social partners. The middle circle represented rather close ones and the outer circle included less close social partners. The SCQ had been put into use recently amongst Hong Kong people from middle-aged to old, which was demonstrated effective to meet the research questions in terms of network structure (Fung, Stoeber, Yeung, & Lang, 2008).

4.4.1.3 Functions of Social Support Network

In a systematic literature review by Tardy (1985), there are three measurement tools designed to explore the functional properties of social support—Arizona Social Support Interview, Inventory of Socially Supportive Behaviors (ISSB) and Social Relationship Scale. Of the three, ISSB is the most consistent with the functional properties of support network in the conceptual framework regarding the present study. Nevertheless, ISSB consists of 40 statements describing the enactment of various content types of support, which was lengthy for interviewing older adults. In addition, ISSB is widely used among college students and adults. Relevant studies among Chinese older adults are sparse.

In view of the past studies of similar topic, measurement instruments of social support function were mostly self-administered tools. Participants were asked to use a Likert rating scale to indicate how helpful each type of support network was in offering the functional support. In this study, emotional support, companionship support, instrumental support and informational support were examined. Accordingly, with reference from the previous empirical studies (Penninx et al., 1997; Wan, Jaccard, & Ramey, 1996), emotional support was described as "listening to you, reassuring you, showing you that they care and willing to share happiness and sorrows". Similarly, instrument support was measured by asking "how often during the previous years do you received support with daily chores around the house (e.g., preparing meals, cleaning the house, transportation, and small repairs) and caregiving". For companionship support, participants were asked "how helpful is each type of network in spending time with you in leisure and recreational activities, in offering a sense of friendship, and in participating in shared activities". Lastly, information support were examined by asking the respondents "how often does each type of network provide information and suggestions on medicine, health, insurance, social welfare, and occupation, etc.?" The types of support network were a list of the previously mentioned five sources of support: primary family members, peer groups, DHs, relatives, and professionals.

4.4.1.4 Emotional Closeness

In the present study, emotional closeness of community-dwelling older adults in Hong Kong was examined to look at its effect on positive aging. Previous research has found that emotional closeness was subject to the perceived psychological closeness between two persons, which might not be necessarily related with the rates of contacts and network size (Carstensen, 1992). Emotional closeness is displayed through the degrees of indicators showing willingness to interact or discuss intimate feelings with the other. However, there were few authorized measures of emotional closeness in empirical studies. Therefore, in the present study, a total of four questions derived from the scale LSNS-R were adopted as the measure of emotional closeness because the selected items were mostly in accordance with the description and definition of emotional closeness. In this measure, two questions were about the closeness with family groups and the other two were about peer groups. Questions were like "When one of your family members and relatives/friends has an important decision to make, how often do they talk to you about it?" and "How often is one of your family members and relatives/friends available for you to talk to when you have an important decision to make?" Respondents were asked to show the intimacy of frequency in terms of the above questions.

4.4.1.5 Positive Aging

Positive Aging Index (PAI) is a scale written in Chinese aiming to comprehensively examine the situation of health, function and engagement with life of Chinese older adults (Ng et al., 2011). It was developed by Ng, Woo, Kwan, and Chong (2006), which was informed by Rowe and Kahn's (1997) construct of successful aging and a qualitative study of positive aging by Chong and associates (2006). Illness avoidance and functioning questions were based on Rowe and Kahn's (1998) construct, with appropriate reference from existing popular scales such as the Activities of Daily Living, Instrumental Activities of Daily Living, and Mini Mental State Examination. With regard to engagement with life, it was divided into two separate themes—CE and PE. CE was based on the support shown to four categories of significant others; each represented one item, namely, family, other relatives, neighbors, and friends. PE was measured by four items relating to productive contributions made to the family, one's work and career, the community, and society (Ng et al., 2011).

However, for the specific study, minor revision was made to the items under CE and PE. Family values and the norm of filial piety have been attached much emphasis from ancient time and still dominate in intergenerational interactions in China. For the assumption that Chinese people normally regard the family as the very important, it is highly recommended to distinguish the implications of CE and limit the caring behaviors into the family context. For PE, the productive contributions were suggested to be embraced into the community and society, excluding the family. Thus, the differentiation of the target audiences of CE and PE was more explicit and apt to Chinese culture. Consequently, CE represented engagement with life which was conducive to family, and PE represented productive behaviors which were beneficial to the community and society.

In light of the results of the pilot study, thematic findings were classified into several main categories in terms of CE and PE respectively. Furthermore, one of the research questions was proposed to identify the ways that older adults engaged with life, thus, it was justified to provide information to the revision of the measurement of CE and PE. Derived from the existing findings of CE in the in-depth interview, the components of CE were presented as household maintenance activities, fostering and educating grandchildren, showing care and support to family members and relatives.

As the theoretical basis of PE, the conceptual *productive aging* has been discussed in social gerontology. Morrow-Howell (2011) re-proposed the conceptual framework of productive aging and broadened the meaning of *productive* comprehensively. Productive aging contained four levels of productive behaviors—activities with economic value, community-building activities, self-care & daily activities, and social & leisure activities. In addition, based on the thematic findings in relation to PE in the in-depth interview, the components of PE were divided into three types of behaviors with economic or social values, namely, volunteerism, paid work, and social & leisure activities.

It should be noticed that there was not a consolidated definition of leisure activities in the previous literature, however, a general identification that contains six cognitive activities, eleven physical activities was accepted in many studies (Verghese et al., 2003; Wang, Karp, Winblad, & Fratiglioni, 2002). In an empirical study conducted in Hong Kong (Chou, Chow, & Chi, 2004), leisure activities of older adults were sorted into seven categories—watching the television or listening to the radio, reading newspapers, books or magazines, socializing with relatives and friends, playing mahjong or cards, doing exercises in the morning or strolling in the park, going out for breakfast and strolling on the street or shopping. While in another study, attending the theater, concerts, or art exhibitions; traveling; playing cards/games; or participating in social groups or a pension organization were regarded as participating in social activities (Wang et al., 2002). Therefore, drawing on the above statement, the overlaps between *social activities* and *leisure activities* could not be avoided in most occasions. Hence, in this study, social and leisure activities represented the activities without the initiative of generating economic value, conceptually including social activities, recreational, artistic and spiritual activities, lifelong learning, civic engagement, entertainment, sports, and so on.

Based on the above re-construct of CE and PE, questionnaire items to explore CE and PE were presented to ask the frequency of participation through a five-point scale ranging from "never", "seldom", "sometimes", "often", "very often".

In addition to the essential elements of PAI—illness avoidance, good functioning, CE and PE, a subjective component to examine older people's subjective well-being was proposed in the framework of positive aging. Specifically, *life satisfaction* was employed in the measurement. It was well-founded to adopt life satisfaction as an element of positive aging since empirical studies had abundant evidence to demonstrate the positive correlation between life satisfaction and perceived well-being,

Table 4.3 Dimensions of the expanded positive aging index (20-item)

Components	No. of items	Descriptions
Illness avoidance	4	General perceived health conditions
Functioning	5	Sleep, memory, basic cognition and functioning
CE	3	Household maintenance, caregiving and education of grandchildren, parenting or support family members and relatives
PE	3	Voluntary work, paid employment, social and leisure activities
Life satisfaction	5	Attitudes towards life

successful aging and positive aging among Hong Kong older adults (Chou & Chi, 1999; Lou, Chi, & Mjelde-Mossey, 2008; Yeung & Fung, 2007). Furthermore, most respondents in the pilot study admitted that embracing content attitudes towards life was beneficial to attain well-being in old age. Considering the limitation of the length of the questionnaire, the Satisfaction with Life Scale (SWLS) was most appropriate as the instrument to examine life satisfaction. It consists of five items of statements describing life satisfaction, using a 1–5 rating scale from "Strongly disagree" to "Strongly agree" (Diener, Emmons, Larsen, & Griffin, 1985).

To conclude, five subscales to measure the corresponding components (i.e., illness avoidance, functioning, CE, PE, and life satisfaction) constitute the 20-item *expanded PAI*, which was set as the measurement of the dependent variable—positive aging. The dimensions and items of the expanded PAI were illustrated as follows (see Table 4.3).

4.4.1.6 Coping

The findings of the pilot study revealed that senior citizens usually adopted several kinds of coping styles if they felt stressful. Generally, coping strategies such as talking with others, asking for others' suggestions, participating in leisure activities, accepting the reality and comforting themselves were most frequently used. Specifically, in light of the empirical studies, it was reported that a series of negative coping strategies (e.g., escape-avoidance, distancing) were proportionally more frequently used among older adults (Folkman, Lazarus, Pimley, & Novacek, 1987).

This study adopted the Simplified Coping Style Questionnaire (SCSQ), a 20-item questionnaire. SCSQ is originally designed for Chinese with the reference of Ways of Coping Questionnaire constructed by Folkman and Lazarus. The Chinese version of this scale has been widely used in mainland (Zhang & Xie, 1995; Hou, Li, Xie, & Zhang, 2004) and Hong Kong Chinese. In Hong Kong, this scale was used in a study to examine the moderating effect of coping on depression among the elders in 2001. The Cronbach's Alpha coefficient in that study was 0.76 (Wong, 2003). Previous studies showed that the SCSQ had satisfactory construct validity (Lou & Mui, 2008).

Table 4.4 Composition of survey questionnaire

Section	Theme and measurement	Question distribution
1	Screening test (SPMSQ)	Q1–Q2, Q3.1–Q3.11
2	Family support network (LSNS-R)	Q4–Q9
3	Peer support network (LSNS-R)	Q11–Q16
4	Emotional closeness	Q8–Q9, Q15–Q16
5	Social network composition (SCQ)	Q18.1–Q18.3
6	Functional properties of social support	Q19–Q22
7	Positive aging (expanded PAI)	Q23–Q34
8	Coping (SCSQ)	Q35.1–Q35.20
9	Demographic variables	Q36–Q55

The scale has a number of 12 items related to active coping strategies and 8 items related to negative coping strategies. Respondents were asked to report the frequency of use of coping strategies (0 = never, 1 = sometimes, 2 = often).

Sum scores were computed for each of the three strategies—problem coping, emotion coping and avoidance coping, the results of the CFA were presented in next section.

4.4.1.7 Summary

Drawing on the descriptions of measurement in term of the variables, the finalized questionnaire of the survey contains nine sections (see Table 4.4).

4.4.2 Scale Validation

In this section, the reliability of construct and the validation of the measurement of positive aging (expanded PAI) and measurement of coping (SCSQ) were introduced. A CFA was conducted on each scale respectively to examine the appropriateness of each factor under the scale.

4.4.2.1 Reliability of Scales and Validation of Expanded PAI

Nunnaly (1978) has indicated 0.7 to be an acceptable reliability coefficient and that lower thresholds sometimes could be used. In light of this, reliability analysis showed that the original composite scales used in the present study exhibited satisfactory internal consistency (see Table 4.5). The Cronbach's Alpha of 20-item expanded PAI was 0.823. The reliability of the scales on family support network and peer support

Table 4.5 Reliability of scales before and after confirmatory factor analysis

Scale	Cronbach's alpha	No. of items	Revised alpha
Positive aging (expanded PAI)	**0.823 (20-item)**		**0.825 (19-item)**
Illness avoidance	0.617	4	
Functioning	0.672	5	
CE	0.816	3	
PE	**0.371**	**3**	**0.531**
Life satisfaction	0.864	5	
Social network composition	0.635	18	
Family support network	0.780	6	
Peer support network	0.758	6	
Functional social support	0.784	20	
Coping (SCSQ)	0.832	20	
Problem coping	0.763	7	
Emotion coping	0.618	6	
Avoidance coping	0.650	7	

network was acceptable with Cronbach's Alpha = 0.780 and 0.758 respectively. In terms of coping, the overall reliability of SCSQ was Cronbach's Alpha = 0.833 and the reliability of its three sub-scales was also acceptable (Cronbach's Alpha = 0.763, 0.618, and 0.650 respectively). Nevertheless, the scale used to measure *PE* under the expanded PAI reflected a low reliability, of which the Cronbach's Alpha was 0.371. Yet, the relatively lower reliability of this composite scale seemed acceptable because we did not assume that its three items would closely relate to each other; instead, the three items of PE measured three forms of productive behaviors, which might not be associated logically or theoretically.

However, it should be noticed that although the measurement of positive aging (PAI) had been previously validated in Chinese society, it was still a relatively new scale. Hence, further application in the present study was required to establish its capability of measuring the new construct with five factors. Therefore, a CFA using IBM SPSS Amos 20 was carried out to further explore the validity of the expanded PAI. Results revealed that all the items could converge into five factors on the condition that the item *undertaking paid work after retirement* was deleted. The CFA results on the original 20-item expanded PAI was illustrated in Fig. 4.2. After deleting that item, the 19-item expanded PAI indicated a relatively satisfactory index of reliability. The Cronbach's Alpha of the revised expanded PAI (19-item) was 0.825. The Cronbach's Alpha of the sub-scale measuring PE (2-item) was 0.531 (see Table 4.5), reflecting a significant improvement.

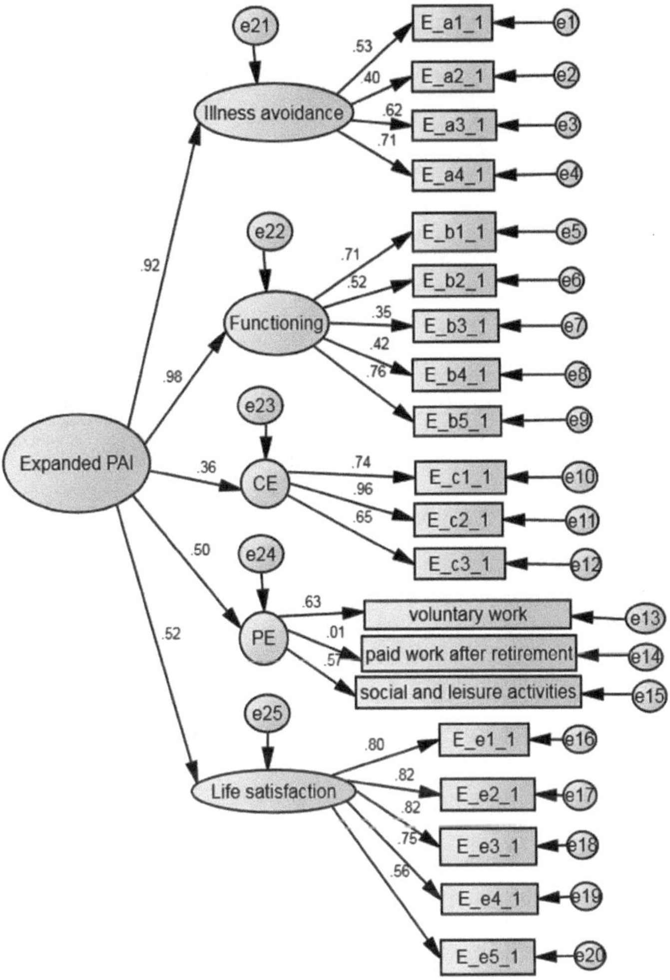

Fig. 4.2 Confirmatory factor analysis of the original expanded PAI (20-item). Note: Goodness-of-fit values: χ^2 (N = 393; *df* = 165) = 312.596; *p* = 0.000; CFI = 0.940; GFI = 0.926; RMSEA = 0.045; SRMR = 0.0559. Standardized β-coefficients were shown and all the coefficients were significant except for the item *undertaking paid work after retirement*

In regard to the changes of model indices before and after deleting that item, the current model reflected slight improvement in terms of comparative fit index (CFI), goodness-of-fit index (GFI), root mean squared error of approximation (RMSEA), and standardized root-mean-square residual (SRMR). CFI was improved from 0.940 to 0.956, GFI was improved from 0.926 to 0.936, RMSEA was modified from 0.048 to 0.043, and SRMR was reduced from 0.0559 to 0.0500 (see Fig. 4.3).

Considering the actual situation in terms of the deleted item *undertaking paid work after retirement*, it is supposed that the respondents in this sample seldom participated

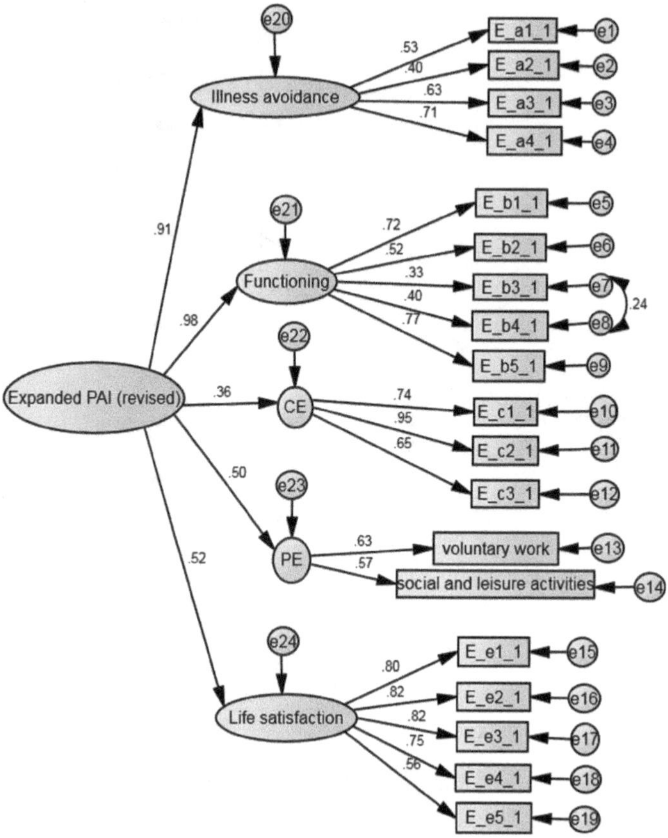

Fig. 4.3 Confirmatory Factor Analysis of the Revised Expanded PAI (19-item). Note: Goodness-of-fit values: χ^2 (N = 393; *df* = 146) = 251.375; *p* = 0.000; CFI = 0.956; GFI = 0.936; RMSEA = 0.043; SRMR = 0.0500. Standardized β-coefficients were shown and all the coefficients were significant

in paid work after retirement. This feature could be resulted from the nature of the sample. The mean age of the sample is 77.3 years old. Respondents reflected that they might have undertaken paid jobs right after their retirement (around 60 years old), when they felt capable to deal with jobs. Finally, the 19-item expanded PAI was adopted in the following steps of data analysis.

4.4.2.2 Validation of SCSQ

Coping in this study was measured by SCSQ. SCSQ is a tool with 20 items in relation to active coping and passive coping. According to the three-type coping strategies (e.g., problem coping, emotion coping and avoidance coping) classified by Ben-Zur (1999), the total 20 items could be categorized into the three meta-strategies.

Table 4.6 Classification of SCSQ in accordance with three-type coping meta-strategies

Type	Sequence No. of items	Statement of items
(1) Problem coping	1	Study or work
	3	See the bright side of things
	4	Change one's thinking
	5	Do not take things too seriously
	6	Stand your ground
	7	Finding several ways out
	9	Change behaviors
(2) Emotion coping	2	Talk with others
	8	Seek advice from relatives, friends and classmates
	10	Learn from others
	11	Actively participate in recreational and sports activities
	12	Emotion control
	17	Rely on others
(3) Avoidance coping	13	Have a rest or holiday
	14	Smoke, drink, take medicine and eat
	15	Wait
	16	Try to forget
	18	Accept the reality
	19	Imagine miracles
	20	Comfort yourself

According to the descriptions and examples of each type of coping strategy, the 20 items were categorized accordingly. Table 4.6 showed the details of classification.

A CFA was conducted to further verify this tool. Results showed that the construct which was composed of three types of coping strategies was tolerably accepted, Chi-square = 437.43, df = 158, p = 0.000, GFI = 0.896, CFI = 0.837, RMSEA = 0.067. Factor loadings of 18 items out of 20 were above 0.30. It is generally accepted the minimum requirement (0.90) for CFI and a moderate range (0.06–0.08) for RMSEA when judging a model fit for CFA (Hu & Benter, 1998). Although the model fit indices seemed not very satisfactory, past evidence argued that CFI $\geq$ 0.80 was acceptable for structural validity of the model (Hai, Anderson, Tatham, & Black, 2009). Additionally, the index of RMSEA = 0.067 in this study was marginally acceptable. Therefore, the results of CFA indicated that the SCSQ with a three-type structure was moderately validated.

4.5 Summary

This chapter presented the methodology of the present study. First, the chapter introduced the mixed-method approach of the two-phase study. The justifications of the methodology were introduced. Second, the research design of the pilot study was presented. The pilot study was an in-depth interview. A number of 16 interviewees were purposely selected from two DECC in two districts of Hong Kong. Thematic findings were recorded and categorized, which provided first-hand information for the scale construction in the second phase of the study. Third, the research design of the main study—survey, was presented. The main study adopted a multi-stage sampling method and obtained a convenience sample of 393 participants of center members in six DECC. Lastly, this chapter introduced the reliability test of scales and validation of the expanded PAI and SCSQ. The CFA was conducted to assess the structural validity of each item under the five factors of expanded PAI. Results of the CFA suggested deleting one item with intolerable factor loading. After deleting the item *undertaking paid work after retirement*, the 19-item was verified as the analytical scale in the following data analysis. Furthermore, the measure of coping (SCSQ) was also validated as the analytical scale which consisted of three types of coping strategies, namely, problem, emotion and avoidance.

References

Babbie, E. R. (2007). *The practice of social research* (11th ed.). Belmont, CA: Thomson Wadsworth.

Babbie, E. R. (2010). *The practice of social research* (12th ed.). Belmont, Calif: Wadsworth Cengage.

Ben-Zur, H. (1999). The effectiveness of coping meta-strategies: Perceived efficiency, emotional correlates and cognitive performance. *Personality and Individual Differences, 26*(5), 923–939.

Braun, V., & Clarke, V. (2006). Using thematic analysis in psychology. *Qualitative Research in Psychology, 3*(2), 77–101.

Carstensen, L. L. (1992). Social and emotional patterns in adulthood: Support for socioemotional selectivity theory. *Psychology and Aging, 7*(3), 331–338.

Census and Statistics Department. (2007). *Summary statistics of district council districts 2006*. Retrieved online from http://www.censtatd.gov.hk/hong_kong_statistics/statistical_tables/index_tc.jsp?charsetID=2&tableID=167&subjectID=1&.

Census and Statistics Department. (2012). *Latest population and household characteristics*. Retrieved online from http://www.censtatd.gov.hk/hong_kong_statistics/dashboard/index_en_GHS.html.

Chi, I., & Boey, K. W. (1992). Validation of measuring instruments of mental health status of the elderly in Hong Kong (Resource Paper Series No. 17). Hong Kong: The University of Hong Kong, Department of Social Work & Social Administration.

Chong, A. M. L., Ng, S. H., Woo, J., & Kwan, A. Y. H. (2006). Positive ageing: The views of middle-aged and older adults in Hong Kong. *Ageing & Society, 26*(2), 243–265.

Chou, K. L., Chow, N. W. S., & Chi, I. (2004). Leisure participation amongst Hong Kong Chinese older adults. *Ageing & Society, 24*(4), 617–629.

Chou, K. L., & Chi, I. (1999). Determinants of life satisfaction in Hong Kong Chinese elderly: A longitudinal study. *Aging & Mental Health, 3*(4), 328–335.

Cochran, W. G. (1977). *Sampling techniques* (3rd ed.). New York: Wiley.

Cohen, J. (1989). *Statistical power analysis for the behavioural sciences* (2nd ed.). Hillsdale, NJ: Erlbaum.

Description of the LSNS. (2011). Bosten college. Retrieved online from http://www.bc.edu/content/bc/schools/gssw/lubben/description_of_thelsns.html.

Diener, E., Emmons, R. A., Larsen, R. J., & Griffin, S. (1985). The satisfaction with life scale. *Journal of Personality Assessment, 49*(1), 71–75.

Folkman, S., Lazarus, R. S., Pimley, S., & Novacek, J. (1987). Age differences in stress and coping processes. *Psychology and Aging, 2*(2), 171–184.

Fung, H. H., Stoeber, F. S., Yeung, D. Y. L., & Lang, F. R. (2008). Cultural specificity of socioemotional selectivity: Age differences in social network composition among Germans and Hong Kong Chinese. *The Journals of Gerontology, Series B: Psychological Sciences and Social Sciences, 63*(3), 156–164.

Hai, J. F., Anderson, R. E., Tatham, R. L., & Black, W. C. (2009). *Multivariate data analysis*. Englewood Cliffs: Prentice Hall.

Hou, S. J., Li, C. G., Xie, Y. N., & Zhang, X. Y. (2004). Investigation of the coping styles and mental health of Military Medical University students. *Chinese Mental Health Journal, 18*, 348–349. (in Chinese).

Hu, L. T., & Bentler, P. M. (1998). Fit indices in covariance structure modeling: Sensitivity to underparameterized model misspecification. *Psychological Methods, 3*(4), 424–453.

Kahn, R. L., & Antonucci, T. C. (1980). Convoys over the life course: Attachment, roles and social support. In P. B. Baltes & O. Brim (Eds.), *Life-span development and behavior* (Vol. 4, pp. 253–286). New York: Academic Press.

Letts, L., Wilkins, S., Law, M., Stewart, D., Bosch, J., & Westmorland, M. (2007). *Guidelines for critical review form: Qualitative studies (Version 2.0)*. Retrieved from http://www.srs-mcmaster.ca/Portals/20/pdf/ebp/qualguidelines_version2.0.pdf.

Li, H. (2010). *Social support and its constructed meaning in the experience of people with physical disabilities: A qualitative study in the city of Changchun in China*. (Unpublished The University of Hong Kong).

Lou, W. Q. V., Chi, I., & Mjelde-Mossey, L. A. (2008). Development and validation of a life satisfaction scale for Chinese elders. *The International Journal of Aging and Human Development, 67*(2), 149–170.

Lou, V.W.Q., & Mui, B. K.W. (2008). *Handbook of measures for the Chinese elderly*. Sau Po Center on Ageing. The University of Hong Kong: Hong Kong.

Lubben, J. E. (1988). Assessing social networks among elderly populations. *Family and Community Health, 11*(3), 42–52.

Marshall, M. N. (1996). Sampling for qualitative research. *Family Practice, 13*(6), 522–526.

Morrow-Howell, N. (2011). *Productive engagement of older adults: Elements of a cross-cultural research agenda*. Conference Presentation in Productive Aging in China: Toward Evidence-based Practice and Policy, Beijing.

Ng, S. H., Cheung, C. K. J., Chong, A. M. L., Woo, J., Kwan, A. Y. H., & Lai, S. F. Y. (2011). Aging well socially through engagement with life: Adapting Rowe and Kahn's model of successful aging to Chinese cultural context. *The International Journal of Aging and Human Development, 73*(4), 313–330.

Ng, S. H., Woo, J., Kwan, A., & Chong, A. (2006). Positive ageing: A conceptual framework. *Korean Journal of Psychological and Social Issues, 12*(5), 29–43.

Nunnaly, J. (1978). *Psychometric theory*. New York: McGraw-Hill.

Peace, S. M., & British Society of Gerontology. (1990). *Researching Social Gerontology: Concepts, Methods and Issues*. London: Sage.

Penninx, B. W. J. H., Van Tilburg, T., Kriegsman, D. M. W., Deeg, D. J. H., Boeke, A. J. P., & van Eijk, J. T. M. (1997). Effects of social support and personal coping resources on mortality in older age: The longitudinal aging study Amsterdam. *American Journal of Epidemiology, 146*(6), 510–519.

Pfeiffer, E. (1975). A short portable mental status questionnaire for the assessment of organic brain deficit in elderly patients. *Journal of the American Geriatrics Society, 23*(10), 433–441.

Population Census, (2012a). Population by sex, marital status, age group and economic activity status, 2011. Retrieved online from http://www.census2011.gov.hk/tc/main-table/C146.html.

Population Census, (2012b). Population by district council district, sex, age group and educational attainment (highest level completed), 2011. Retrieved online from http://www.census2011.gov.hk/en/main-table/B209.html.

Rowe, J. W., & Kahn, R. L. (1997). Successful aging. *The Gerontologist, 37*(4), 433–440.

Rowe, J. W., & Kahn, R. L. (1998). *Successful aging*. New York: Pantheon.

Steckler, A., McLeroy, K. R., Goodman, R. M., Bird, S. T., & McCormick, L. (1992). Toward integrating qualitative and quantitative methods: An introduction. *Health Education Quarterly, 19*(1), 1–8.

Tardy, C. H. (1985). Social support measurement. *American Journal of Community Psychology, 13*(2), 187–202.

Tashakkori, A., & Teddlie, C. (2003). *Handbook of mixed methods in social & behavioral research*. Thousand Oaks, California: SAGE Publications.

Teddlie, C., & Yu, F. (2007). Mixed methods sampling. *Journal of Mixed Methods Research, 1*(1), 77–100.

Verghese, J., Lipton, R. B., Katz, M. J., Hall, C. B., Derby, C. A., Kuslansky, G., et al. (2003). Leisure activities and the risk of dementia in the elderly. *New England Journal of Medicine, 348*(25), 2508–2516.

Wan, C. K., Jaccard, J., & Ramey, S. L. (1996). The relationship between social support and life satisfaction as a function of family structure. *Journal of Marriage and Family, 58*(2), 502–513.

Wang, H. X., Karp, A., Winblad, B., & Fratiglioni, L. (2002). Late-life engagement in social and leisure activities is associated with a decreased risk of dementia: A longitudinal study from the Kungsholmen project. *American Journal of Epidemiology, 155*(12), 1081–1087.

Wong, B. (2003). *The moderating effect of coping on depression in the elderly: A Hong Kong study* (Unpublished M.Phil. thesis). The University of Hong Kong.

Yeung, G. T. Y., & Fung, H. H. (2007). Social support and life satisfaction among Hong Kong Chinese older adults: Family first? *European Journal of Ageing, 4*(4), 219–227.

Zhang, Y. K., & Xie, Y. N. (1995). Correlation analysis of mental health, social support and coping style with minority undergraduates. *Chinese Journal of Clinical Psychology, 3*, 73–74. (in Chinese).

Chapter 5
Findings of In-Depth Interview

The pilot study was successfully conducted in the two selected districts of Hong Kong in March and April, 2012. With the help of the colleagues in the District Elderly Center of the Caritas (Sham Shui Po) and the Wunsha Elderly Center of St. James' Settlement, a total of 16 older people were invited to participate in the in-depth interview. For those who had good functioning abilities, interviews were performed in the case interview room of the institute. As for the home-bound older adults, home visits were conducted. In the process of individual interview, interviewees were asked to talk about freely according to the questions from the interview guidelines. All the dialogues were taped and then transcribed verbally. Thematic analysis was adopted to categorize all the significant findings into main themes in light of the proposed conceptual framework.

5.1 Case Profiles

The detailed profiles of sixteen interviewees in total are as follows (see Table 5.1). From table, nine men and seven women participated in the interview. The age of participants ranged from 71 to 90 and the mean age was 80.25. A number of five participants reported they were self-care incapable. In terms of the situation of co-inhabitance, ten participants said they lived with their spouse, three lived with the daughter and three were alone. For the perceived SES, five respondents reported in high level of SES, six in middle level and the other five in low level.

The case profiles of the in-depth interview reflected that the sample had an even distribution of gender, age, and perceived SES.

S. Liu, *Social Support Networks, Coping and Positive Aging Among the Community-Dwelling Elderly in Hong Kong*,
https://doi.org/10.1007/978-981-13-3783-3_5

Table 5.1 Case profiles of in-depth interview

Case No.	Gender	Age	Location	Health	Co-inhabitance	Perceived SES
C-1	M	72	SSP	Self-care capable	Spouse	Middle
C-2	F	85	SSP	Self-care incapable	Daughter	High
C-3	F	77	SSP	Self-care capable	Daughter	High
C-4	F	80	SSP	Self-care capable	Spouse	Middle
C-5	F	78	SSP	Self-care capable	Daughter	Middle
C-6	M	78	SSP	Self-care capable	Spouse	High
C-7	M	85	SSP	Self-care capable	Spouse	Middle
C-8	M	87	SSP	Self-care incapable	Alone	Low
C-9	M	72	WC	Self-care capable	Alone	Middle
C-10	F	71	WC	Self-care capable	Spouse	High
C-11	M	79	WC	Self-care capable	Spouse	High
C-12	M	89	WC	Self-care capable	Alone	Middle
C-13	M	90	WC	Self-care incapable	Spouse	Low
C-14	F	85	WC	Self-care incapable	Spouse	Low
C-15	F	78	WC	Self-care incapable	Spouse	Low
C-16	M	78	WC	Self-care capable	Spouse	Low

Note SSP represents Sham Shui Po District; WC represents Wan Chai District

5.2 Thematic Findings

In this section, thematic findings on the main variables—social support network, positive aging, and coping were summarized and presented. Additionally, participants' attitudes of stress in later life from participants were included.

5.2.1 Thematic Findings on Social Support Network

In terms of social support, respondents reported that spouse and adult children were the major resources to provide support (e.g., financial support, emotional support). Nevertheless, all the older parents expressed their understanding and tolerance to their adult children for not having enough time to stay with them. They fully understood that the children had their own career and should struggle for a living. Therefore, the old parents did not lay too many expectations on children and seldom asked for their monetary assistance.

In addition to the two major types of social support providers, interviewees said they were more inclined to keep in touch with the old friends, such as colleagues and workmates, as long as they were alive and at leisure. However, many old friends were living within a distance from the interviewees so that they could not meet regularly. Thus, as a matter of fact, older people did not receive much emotional support and companionship support from their old friends. For the older people in poor conditions, they preferred to receive instrumental support from their spouses, if the spouses could afford. For older adults living in the community, they got companionship support from the peer groups who were also members of community center. They were fond of chatting and playing with the peer groups and regarded this as part of their daily life. Few findings were discovered in terms of the role of domestic helper in providing support to the old. None of the respondents has the idea to employ a full-time domestic helper to help household affairs or take care of the home-bound older persons. The below quotes were extracted from the transcripts.

5.2.1.1 Family Support

I do not do extra work at weekends and I want to stay with my wife. (C-11)

I have serious neurasthenia and many kinds of physical deterioration, as well as hearing impairment. So my wife keeps taking care of me for many years and I can't live without her. (C-16)

Parents are always thinking about their children, including younger generation. I know my children are busy and have their own career; they do not come to see us very often. (C-13)

My son is engaged with his work and social activities, and he likes drinking. So he doesn't often stay with me. But I know it's his life style. So does my adopted daughter. We are ready to help each other if necessary. (C-11)

I have two sons residing in Australia. Another son is in Hong Kong, but we are not in a good relationship, so never get in touch with each other. I have my own life style and seldom miss them. (C-9)

My wife and I love to play with my little grandson, he's so cute. (C-6)

At that time I had a very severe heart disease and entered ICU for a couple of days. It's my wife who provided continuous caregiving to me. (C-9)

I felt my health was not as good as before when I was 83 years old. My husband is my caregiver. (C-14)

5.2.1.2 Peer Support

There is a policeman union organized for the retired policemen. So we often have activities on festivals and public holidays. Also, the neighborhood welfare institute also holds interesting activities for us. (C-12)

I often play with my previous colleagues and workmates, for example, having dinner and playing Mahjong. (C-11)

I like to chat with the members in this center. I often comfort those members who are worried about that their children seldom come to see them. I like to chat with the members in this center. (C-4)

I am not sure whether a foreign domestic helper could speak Cantonese because I know many Filipino could not speak Cantonese and they are not smart at all in caregiving. (C-14)

If I have some worries I will chat with the peer members. I seldom talk about them with the social workers here. (C-2)

I live alone and have no children. I still keep in touch with my old colleagues and friends. Sometimes we play together. (C-8)

Now I am content with the current life as long as my husband could accompany me. (C-14)

Now very often my previous colleagues and friends invite me to have tea, go to the theater and sing songs together. I feel so happy to join in these activities. (C-10)

For my previous colleagues and friends, we have less contact than before because we have already retired and they might not have leisure time. (C-6)

I keep in touch with few workmates now. (C-15)

5.2.2 *Thematic Findings on Positive Aging*

Among the findings related with positive aging dimensions, all the respondents recognized the importance of physical health, and many of them attempted to control illness deterioration. Facing the fact that some chronic diseases were probably hard to cure, they accepted them and did not take them so seriously. Of all the respondents, only three of them had severe functioning problems, the others had good functioning abilities. In regard to caring engagement, most of the respondents reported that they could shoulder the basic housework at home, including cooking, washing, and cleaning. Several of them had helped foster grandchildren before. A majority of the sixteen interviewees have experiences working as volunteers, among which, many of them really loved to make continued contributions to others by serving as volunteers. A couple of respondents said they continued to work for a few years after retirement for the reason that they were in good health. Moreover, it would be better to earn money as savings for later life. Most of the older adults were content with their current life and generally they held a lower or realistic expectation on their future life. The below quotes were extracted.

5.2.2.1 Illness Avoidance

Now I do not have any diseases at all. Maybe it could be attributed to my interests in sports such as playing basketball and table tennis from my younger time till now. (C-3)

At current time I do not have an obvious feeling of growing old. The most impressed point is that I would recover so soon at my younger age; however, it seems a bit longer for me to recover if I feel tired or uncomfortable. However, actually I don't have any other serious diseases except osteoporosis. (C-5)

5.2.2.2 Functioning

Although I have a little diabetes, I do not have the feeling of becoming old. Because I walk fast and handle affairs very quickly. (C-10)

I can bathe myself and do some simple exercises at home. I can do some simple cooking myself. (C-2)

5.2.2.3 Caring Engagement

I have to take care of my ill husband and his very old mother. I do almost all the housework and cook for ourselves. (C-15)

I have fostered my elder granddaughter. (C-14)

In weekdays I do grocery shopping and cook at home, watch TV or hang around the streets. (C-13)

5.2.2.4 Productive Engagement

I have a lot of things to do in this center, such as organizing meetings, doing some office affairs, helping measure blood pressure and joining debate teams, which are so interesting. (C-7)

I have a lot of things to do in this center, such as organizing meetings, doing some office affairs, helping measure blood pressure and joining debate teams, which are so interesting. (C-6)

For the political election, I have voted once or twice. Sometimes I pay attention to what they have done. (C-4)

5.2.2.5 Life Satisfaction

I do not have any extravagant hopes on life because I know there are a lot of people whose livelihoods are poorer than me. (C-5)

I don't have any high expectations on life. I think one should be content with his life and maintain a happy mood. There is no need to concern and worry about others too much. (C-15)

5.2.3 Thematic Findings on Stress

Almost half of the respondents reported that they had a strong feeling of becoming old in a couple of years after retirement. The major symptoms were the deterioration of memory, appearance of chronic diseases and body aches, etc. However, most of the older adults did not regard this as a severe stress and peacefully accepted it as a normal process. But as to the older people suffering many kinds of diseases and poor functioning, they reflected some worries about this. In addition to the threats of physical health, a few reported that they were still worried about their children's living. It is because that some of the adult children were struggling for life and had to shoulder the responsibility of their own nuclear families in the current stagnant economy in Hong Kong. On the contrary, a vast majority of the interviewees indicated no worries about their own life (e.g., the financial resource); instead, all of them did not have any luxury hopes in life. Additionally, several respondents said they felt depressed during the period of time when their spouse passed away.

I felt my health was not as good as before when I retired. I have hypertension and a high Uric Acid. (C-6)

At present I am a bit worried about the relationship between my son and daughter-in-law. They are not communicating so well, which makes me a little anxious. But in many occasions I could not do anything to improve, because it is their personal affair. Sometimes I have insomnia. (C-10)

I have too many diseases, such as diabetes, hypertension, high cholesterol and stroke. I have suffered a lot from the above diseases and in many occasions it makes me feel depressed. (C-15)

5.2.4 Thematic Findings on Coping

Confronted with the hassles and stressors in life, most of the respondents indicated that they had their own ways to cope with them. For the worries related with health problems, all of the respondents were aware of a few positive available approaches to keep fit and maintain a good physical condition. For example, they did exercises regularly, took medicine accordingly and did some aerobic exercises properly (e.g., jogging, playing basketball, dancing). Additionally, for those who suffered from chronic diseases and functional deterioration, from which they might not recover, on the whole, the respondents reflected some negative coping strategies such as trying to accept the reality and comforting themselves. Remarkably, a majority of the respondents showed an obvious characteristic of independency. It may probably point to a belief that individuals should rely on themselves and have their own plans on what they want to do, not dependent on others' help and sympathy. When asking these people about how to cope with difficulties and stressors, they generally reported a series of positive coping strategies, for instance, studying, insisting on personal standpoint, endeavor to figure out methods to solve problems.

Another finding in terms of coping was negative/avoidance coping styles, which was more prevalent among the interviewees who were religious. Some participants would like to adopt some avoidant attitudes towards stressors (e.g., letting go, comforting oneself) in the circumstances where problems were probably impossible to be solved.

I've never thought about I'm in my old age now. I can do what others can do. If I want to do something, I'll think over and try to make it. (C-4)

While I did not give myself too much pressure because I knew death was within a blink of eyes for those who suffer from heart diseases. Thereby I try to make myself happy and usually tell jokes with others. (C-9)

I live with my wife in his public house. Actually I am not physically healthy. I have set up a cardiac pacemaker for 28 years. And my foot tendon was accidently damaged by a doctor because of his mistake several years ago. I do not have much money but I have accepted the reality. (C-13)

I am not active in other activities. I just watch TV at home. I am so quiet. (C-16)

5.3 Summary

This chapter presented the thematic findings in the first phase of the study. The findings of the in-depth interview collected first-hand information in terms of social support network, positive aging, stress and coping. Qualitative findings confirmed the important role of primary family members (e.g., spouse, adult children) in providing emotional and instrumental support. Furthermore, the respondents reported the companionship support provided by peer groups (e.g., old colleagues, center members). Participants also stated that they did not lay many expectations on adult children for the reason that they need engage in career and their own family. Therefore, they would be very happy if the adult children gave support to them. In regard to the components of positive aging, results showed that participants agreed on the importance of the five components, which were essential in contributing to positive aging. As to stress and coping, participants expressed that they accepted the health and functioning declines occurred in later life because aging was irreversible. Furthermore, participants reported they still concerned and worried about their primary family members (e.g., spouse, children and grandchildren) very much. The findings showed that a content attitude towards the current life, a positive mindset and a sometimes "letting go" notion were good recipes to cope with stressors.

The findings of the in-depth interview collected substantial data related to the research focus of the study and provided adequate information in the process of scale construction and modification in the main study.

Chapter 6
Descriptive Findings of Survey

In this chapter, simple descriptive statistics of the survey are presented to examine the participants' profile and to provide an overview of the study. This chapter consists of four parts. Part one presents the participants' profile in the survey and their demographic information. Part two displays the descriptive statistics of the respondents in terms of the main variable in this study. Part three exhibits the correlations among main variables. The last part is the summary of the whole chapter.

6.1 Sample Characteristics

Short Portable Mental Status Questionnaire (SPMSQ) was used at the beginning of the interview to screen for eligibility in cognitive functioning. Finally, questionnaires from a total of 393 older adults aged 65 or above who lived in the community were obtained in the survey. In this sample, 62 participants were from Ma On Shan DECC (15.8%), 67 were from Chan Tseng Hsi DECC (17.0%), 60 were from Ho Kin DECC (15.3%), 68 were from St James Settlement DECC (17.3%), 67 were from Holy Carpenter Church DECC (17.0%), and the last 69 were from Lok Man Alice Kwok Integrated Service Center (17.6%). The participants are distributed among Hong Kong Island, Kowloon and New Territories.

6.1.1 Demographic Factors and the Sample Features

The current survey failed to use random sampling method, resulting in a non-probability sample of Hong Kong older Chinese living in the community. The sample included 101 men and 292 women with a mean age of 77.3 (*SD* = 7.3). The average

S. Liu, *Social Support Networks, Coping and Positive Aging Among the Community-Dwelling Elderly in Hong Kong*,
https://doi.org/10.1007/978-981-13-3783-3_6

participant had primary school education (*M* = 2.6, *SD* = 1.4). Of all the participants, 33.8% were married with their spouse alive. The mean monthly personal income (including wage, pension, old age allowance, living allowance from others, and so on) was below HKD 2000 (*SD* = 1.1). In terms of perceived SES, the mean was reported to be lower-middle (*SD* = 0.9). Though the representativeness of the present sample was not strong because of the nature of sampling method, a comparison between the present sample and a random sample of Hong Kong Household Survey (2009) was conducted to look at the similarities and diversities of two samples. The respondents' profile of this survey and relevant profile of the Household Survey (2009) were reported in Table 6.1.

From the profile of respondents of the survey, it could be found that female account for 74.3% of the total sample. About one third of the respondents were illiterate. Over half of the participants have received elementary or junior education. In terms of the marital status, over half of the participants (53.7%) were widowed. The rate of older adults who lived alone in this sample was about 43.8%, nearly accounting for half of the total sample. The proportion of living with spouse or children was 39.2%. The percentage of living with persons other than spouse and children was 17.0.

In comparison with the situation of Hong Kong community-dwelling elderly, the present sample shared some similarities in terms of age distribution, education and household composition, etc. The target group of the present study was the elderly aged 65 and above. In addition to the group 60–64 and the group aged over 75, the present sample was close to the age distribution of Hong Kong in terms of the older adults aged 65–69 and 70–74. Additionally, the situation of educational attainment in the present sample was very similar with the Hong Kong population, except for those who did not attend school before. In terms of household composition, the proportion of living with children and living with spouse in the present study were close to the Hong Kong population.

The present sample reflected its unique characteristics due to the sampling criteria. The likely reason of the large difference in the number of men and women is that Chinese women generally have a longer life expectancy and used to go to community center more frequently. A report of the Executive Summary of District Elderly Community Center (DECC) and Neighborhood Elderly Center (NEC) (Social Welfare Department, 2006) said that the proportion of women members accounted for a larger share (70%) than that of the total population of Hong Kong (52%). What's more, the proportion (51%) of center members who had not received schooling was higher than the total population (28%). Additionally, the high proportion of widowed respondents may be due to the fact that widowed elderly were more likely to attend community center compared with those with a spouse at home.

Two specific features of the sample need to be addressed. First, the present sample did not cover all types of service users of DECC. The participants were a portion of members who came to the center on a regular basis. Generally, DECC provides a series of services that include community education, case management, reaching out and networking, health education, volunteer development, carer support services, social and recreational activities, meal and laundry services, etc. (Social Welfare

Table 6.1 Demographic Variables Comparison between the Present Sample and Household Survey (2009)

Demographic variables	Present sample (frequency/percentage)	Hong Kong Community-dwelling Elderly[a] ('000) (frequency/percentage)
Age group		
60–64	–	301.7 (26.7%)
65–69	68 (17.3%)	221.5 (19.6%)
70–74	71 (18.1%)	230.5 (20.4%)
≥75	254 (64.6%)	376.3 (33.3%)
Gender		
Male	101 (23.7%)	546 (48.3%)
Female	292 (74.3%)	584 (51.7%)
Marital status		
Unmarried	21 (5.3%)	31.5 (2.8%)
Separate/divorced	28 (7.2%)	52.8 (4.7%)
Married	133 (33.8%)	762.4 (67.4%)
Widowed	211 (53.7%)	283.3 (25.1%)
Education		
No schooling and pre-primary	166 (42.2%)	325.9 (28.8%)
Primary	123 (31.3%)	426.0 (37.7%)
Secondary/sixth-form	78 (19.9%)	300.3 (26.6%)
Post-secondary	26 (6.6%)	77.7 (6.9%)
Household composition		
Living alone	172 (43.8%)	143.5 (12.7%)
Living with spouse[b]	62 (15.8%)	278.8 (24.7%)
Living with children[b]	63 (16.0%)	223.7 (19.8%)
Living with spouse and children[b]	29 (7.4%)	443.6 (39.3%)
Living with persons other than spouse and children	67 (17.0%)	40.3 (3.5%)

Note [a]Hong Kong Community-dwelling Elderly population is inferred from the random sample of Hong Kong Thematic Household Survey in terms of community-dwelling elderly (2009). Frequency is calculated in thousand

[b]Including those older persons living together with spouse/children **and** other persons

Department, 2014). For the present sample, most of the participants were the users of social and recreational activities, meal and laundry services and community education. Second, the sample of the survey encompassed a majority of older people with relatively lower SES. According to the report of the Executive Summary of District Elderly Community Center (DECC) and Neighborhood Elderly Center (NEC) (Social Welfare Department, 2006), DECC members might contain a larger percentage of elders living in public rental housing (39% in HK vs. 52–53% among NECs/DECCs members).

It should be noticed that the non-probability sample in this study cannot lead to generalized conclusion of the findings. Although the similarities between the current sample and the sample of Hong Kong Household Survey could not be ignored, the findings of the current research should be interpreted with caution by taking account of the specificity of the sample.

6.1.2 Social-Economic Factors and Indicators of Elder Caregiving

In this section, social-economic factors reflecting the elderly's socioeconomic status (SES) and indicators reflecting the physiological needs of caregiving are presented in Table 6.2. Results showed that over 60% of the respondents come from the lower and lower-middle class. It was regarded as a unique characteristic of this sample. About 15.5% of the participants have reported good health without any type of diseases. Participants suffering 1 to 2 diseases and 3 to 4 diseases accounted for 55% and 20.8% respectively. About 5.9% of older adults suffered more than 5 types of diseases. In terms of the primary caregivers of the elderly, 37.7% of the respondents reported they did not need caring at home. Additionally, spouse and children (including children-in-law) shouldered the tasks of caregiving of the older family members, which accounted for 18.1% and 38.7% of the sample respectively. During the interview, participants were also asked their willingness of moving to residential care homes when their health deteriorated. Among all the respondents, 41% said they would like to stay in institutions in the future if possible. In addition, 38.9 and 19.3% of the respondents reported refusal and uncertainty to stay in institutions, respectively.

Descriptive statistics revealed that the participants in this sample were mainly in lower and middle socioeconomic status and in moderately good health. These features may result from the nature of the sample that the participants were recruited from the DECC.

Table 6.2 Social-economic factors and indicators of elder caregiving (N = 393)

Variables	Scoring or Range	Frequency (Percentage %)	Mean (SD)
Perceived SES	1–5		1.96 (0.876)
Lower class		144 (36.6)	
Lower-middle class		127 (32.3)	
Middle class		107 (27.2)	
Middle-upper class		7 (1.8)	
Upper class		2 (0.5)	
Variety of chronic diseases incidence	0–11		1.80 (1.53)
0		61 (15.5)	
1–2		216 (55)	
3–4		82 (20.8)	
≥5		23 (5.9)	
Primary caregiver	NA		NA
None[a]		148 (37.7)	
Spouse		71 (18.1)	
Children and children-in-law		152 (38.7)	
Others		19 (4.8)	
Willingness to stay in residential care home	1–3		1.98 (0.898)
Yes		161 (41.0)	
Not clear		76 (19.3)	
No		153 (38.9)	

Note None[a] in terms of primary caregiver stands for no need for caregiving

6.2 Descriptive Statistics of Main Variables

This section mainly displayed the descriptive statistics of the main variables in the survey. The mean and standard deviation of main variables were presented in Table 6.3. As the table showed, the reported total SNS had a wide range from 0 to 76, with a mean size of 10.2 ($SD = 9.4$). The size of close social network and peripheral social network had a range of 0–31 and 0–62 respectively, with a mean size of 3.64 ($SD = 3.95$) and 6.57 ($SD = 8.05$). Moreover, the size of family network (6–36) and that of peer network (6–34) were similar. T-test showed that the difference between family network size ($M = 18.09$, $SD = 6.39$) and peer network size ($M = 18.60$, $SD = 6.08$) was not significant, $t\ (383) = -0.131$, $p \geq 0.05$. However, it should be noticed that this ambiguous finding might be due to lacking of statistical power. To further examine this issue, a power analysis was performed via PASS. According to the statistics, there were a number of 69,716 DECC members in the 2012–2013 cohort (Hong Kong Annual Digest of Statistics, 2013). The result of paired means

Table 6.3 Mean and std. deviations of main variables of the survey (N = 393)

Main variables	Scoring (range)	Mean	Std. deviation
Social support network	0–76	10.2	9.4
Size of close social network	0–31	3.64	3.95
Size of peripheral social network	0–62	6.57	8.05
Primary network size * t (383) = −0.131, $p = 0.184$			
Family network size	6–36	18.09	6.39
Peer network size	6–34	18.60	6.08
Primary functional support * t (391) = 9.866, $p = 0.000$			
Functional family support	1–5(1–5)	2.67	1.07
Functional peer support	1–5(1–4.75)	2.15	0.81
Emotional closeness * t (390) = 8.638, $p = 0.000$			
Family emotional closeness	1–6 (1–6)	2.91	1.48
Peer emotional closeness	1–6 (1–6)	2.21	1.23
Coping* F(1.95, 719.19) = 19.07,$p = 0.000$	0–2 (0–1.65)	0.74	0.35
Problem coping	0–2 (0–1.86)	0.75	0.48
Emotion coping	0–2 (0–1.83)	0.66	0.41
Avoidance coping	0–2 (0–1.71)	0.79	0.40
Positive aging	1–5(1.57–4.71)	3.14	0.67
Illness avoidance	1–5(1.5–5.0)	3.43	0.84
Functioning	1–5(1.4–5.0)	3.46	0.79
Caring engagement (CE)	1–5 (1.0–5.0)	2.51	1.26
Productive engagement (PE)	1–5(1.0–5.0)	2.64	1.22
Life satisfaction	1–5(1.0–5.0)	3.50	0.74

Note variables with a * mean that a T-test or One-way ANOVA analysis of variances have been conducted to compare the differences between the means. Results of the test were presented

power analysis showed that in line with the present evidence, a minimum sample size of 1659 from a population of 69,716 could achieve 80% power to detect a mean of paired differences of 0.5 with a known standard deviation of differences of 7.4 and with a significance level (alpha) of 0.05 using a two-sided paired z-test. Thus, it could be inferred that the current sample size might not endow adequate power to detect the mean difference between family network size and peer network size.

In terms of the functional support, functional family support had higher scores ($M = 2.67$, $SD = 1.07$) than functional peer support ($M = 2.15$, $SD = 0.81$). T-test showed that the difference between functional family support and functional peer support was significant, t (391) = 9.866, $p < 0.00$. With regard to the situation of emotional closeness, emotional closeness with family members had a mean score of 2.91 ($SD = 1.48$) and emotional closeness with peer group had a mean score of 2.21 ($SD = 1.23$). T-test showed that the difference between the two was significant, t (390) = 8.638, $p < 0.00$.

As to the coping strategies in this study, the average score of coping (SCSQ) was 0.74 (*SD* = 0.35) with a score ranging from 0 to 1.65. Statistics also presented that participants had a prevalent use of avoidance coping (*M* = 0.79, *SD* = 0.40), followed by problem coping (*M* = 0.75, *SD* = 0.48) and emotion coping (*M* = 0.66, *SD* = 0.41). A one-way repeated measures ANOVA showed that a significant difference among the three coping strategies, $F\,(1.95, 719.19) = 19.07, p < 0.000$. Meanwhile, it should be interpreted that the frequency of use of the coping strategies was less than "sometimes". For the overall positive aging and its components, participants had an overall score of positive aging at 3.14 (*SD* = 0.67), of which the range of score was from 1.57 to 4.71. Among the components, the mean scores of illness avoidance (*M* = 3.43, *SD* = 0.84) and functioning (*M* = 3.46, *SD* = 0.79) were similar, while PE (*M* = 2.64, *SD* = 1.22) embraced a higher score than CE (*M* = 2.51, *SD* = 1.26). In terms of the component of life satisfaction, the mean score was 3.50 (*SD* = 0.74), representing the general attitudes towards life satisfaction in later life was "slightly agree".

6.3 Correlations Between Main Variables and Demographic Factors

Before examining the predictive effect of social support network on positive aging, correlation analysis were conducted to provide a preliminary review of the relations among the main variables under scrutiny. These included main dependent variables—overall positive aging and its five components (e.g., illness avoidance, functioning, CE, PE and life satisfaction), main independent variables—social support network (e.g., size and function), emotional closeness (e.g., family and peer emotional closeness), and the mediator—coping (e.g., problem, emotion, and avoidance).

The correlations between independent variables, demographic variables and dependent variables were shown in Table 6.4. The results of the correlation between demographic factors (e.g., age, gender, education and perceived SES) and dependent variables were also presented. Results showed that age was negatively associated with positive aging and its components except for life satisfaction. In terms of gender, it revealed that men tended to have better physical and functioning health than women, while women seemed to have more productive engagement than men. Education and perceived SES were positively correlated with all the components of positive aging, reflecting that higher education background and socio-economic status were associated with better achievement of positive aging.

Moreover, in terms of the correlations between main independent variables and dependent variables, social network size and family network size were positively correlated with positive aging respectively except for illness avoidance, while peer network size was positively associated with all the components of positive aging. Functional social support provided by family members was positively associated

Table 6.4 Correlations between demographic variables, independent variables and dependent variables (N = 393)

Demographic and independent variables	Positive aging					
	IA	F	CE	PE	LS	Overall PA
Age	−0.228**	−0.247**	−0.225**	−0.215**	−0.023	−0.275**
Gender	−0.116*	−0.160**	0.010	0.105*	0.035	−0.053
Education	0.220**	0.200**	0.222**	0.106*	0.119*	0.220**
Perceived SES	0.106*	0.124*	0.191**	0.122*	0.346**	0.283**
Social network size	0.095	0.172**	0.187**	0.254**	0.184**	0.272**
Family network size	0.037	0.121*	0.439**	0.289**	0.269**	0.334**
Peer network size	0.143**	0.191**	0.253**	0.253**	0.217**	0.314**
Functional family support	0.018	0.073	0.371**	0.191**	0.243**	0.268**
Functional peer support	0.123*	0.175**	0.254**	0.255**	0.238**	0.318**
Size of close network	0.032	0.028	0.036	0.044	0.190**	0.130*
Size of peripheral network	0.096	0.188**	0.201**	0.276**	0.121*	0.257**
Family emotional closeness	0.068	0.158**	0.378**	0.264**	0.252**	0.326**
Peer emotional closeness	0.076	0.138**	0.215**	0.167**	0.134**	0.212**
Problem coping	0.175**	0.300**	0.275**	0.370**	0.186**	0.386**
Emotion coping	0.096	0.220**	0.321**	0.367**	0.116*	0.328**
Avoidance coping	0.071	0.137**	0.090	0.145**	−0.018	0.130*

Note Gender: 1 = Male, 2 = Female. IA = Illness Avoidance, F = Functioning, CE = Caring Engagement, PE = Productive Engagement, LS = Life Satisfaction. Overall PA = Overall Positive Aging
*$p < 0.05$, **$p < 0.01$

with the components of CE, PE, LS and overall positive aging. However, functional support from peer group was positively correlated with all the components of positive aging as well as the overall score. The size of close network was only positively associated with life satisfaction and overall positive aging, while the size of peripheral network was positive associated with the components of positive aging except for illness avoidance. In terms of emotional closeness, both family and peer emotional closeness were positively associated with the components of positive aging respectively except for illness avoidance. As to the three types of coping strategies, findings revealed that problem coping was positively associated with all the components as well as overall positive aging. Emotion coping was positively correlated with other four components except illness avoidance. However, in terms of avoidance coping, it was only positively correlated with functioning, PE and overall positive aging.

In summary, the listed demographic factors and main independent variables were significantly correlated with the overall positive aging except gender. Yet, it should be noted that the correlation analysis could only provide a preliminary indication of the relationships among the main variables. The relationships revealed by this analysis might be spurious. In any case, the important correlational findings between variables provide solid and fundamental foundation for the inferential analysis later on. Additionally, the insignificant relationship between illness avoidance and other variables may be due to the nature of health condition of older adults, which is deemed as independent to psychosocial factors.

6.4 Summary

In this Chapter, the descriptive findings of the survey were presented. This chapter first displayed the demographic information of the sample. For the reason that the present survey did not adopt a random sample, a comparison between the current sample and representative Hong Kong household sample was conducted. It could be derived from the comparison that the current sample shared a variety of common characteristics, though the diversities on the proportion of widowed elderly, women and the old-old could not be ignored. Second, the chapter presented the descriptive findings of main variables including the range, mean and standard deviation. Third, the correlation analysis findings were displayed. It revealed that a majority of the selected variables were significantly correlated with the components of positive aging as well as overall positive aging score.

References

Hong Kong Annual Digest of Statistic. (2013). *Census and statistics department*. Retrieved online at www.censtatd.gov.hk.

Social Welfare Department. (2006). *Executive summary of DECC and NEC*. Retrieved online from www.swd.gov.hk/doc/.../ExecutiveSummary_chi.doc.

Social Welfare Department. (2014). *Opportunities for the elderly project*. Retrieved online from http://www.swd.gov.hk/en/index/site_pubsvc/page_elderly/sub_csselderly/id_opportunit/.

Thematic Household Survey Report No. 40. (2009, June 4). *Socio-demographic profile, health status and self-care capability of older persons, census and statistics department*. Retrieved online at http://www.statistics.gov.hk/pub/B11302402009XXXXB0100.pdf.

Chapter 7
Hypotheses Testing

To examine the research questions and research hypotheses proposed in previous chapters, this section presents the results of data analysis. In the process of data analysis, IBM SPSS 21.0 and AMOS20.0 were used to test the hypotheses. The sequence of hypothesis testing was based on the hypotheses listed in Chap. 3 (see Sect. 3.3).

7.1 Hypotheses Testing

7.1.1 Testing of Hypothesis Set 1

Hypothesis set 1 examined the structural properties (e.g., size) and functional properties (e.g., the sum of instrumental, emotional, companionship and informational support) in terms of close social networks, peripheral social networks, family networks and peer networks. Their predicting effects on each single component and on the overall positive aging were tested. In addition, the effect of emotional closeness (i.e., family and peer emotional closeness) on positive aging was examined.

To test Hypothesis 1.1–1.5 (see Table 7.1), hierarchical linear regression analysis was conducted on the basis that previous descriptive findings revealed the correlations between variables in regard to family network, peer network, social network size and positive aging. In the process of linear regression analysis, each component and the overall positive aging were set as the dependent variables. Then, age, gender, education and perceived SES were entered in step 1 as the covariates. The independent variables were entered in step 2 in each analytical model, respectively.

Results of the testing of Hypotheses 1.1–1.3 through the hierarchical linear regression analysis were shown in Table 7.2. Statistics showed that family network size ($\beta = 0.169$, $p < 0.001$), peer network size ($\beta = 0.190$, $p < 0.001$) and social network size ($\beta = 0.133$, $p < 0.01$) all positively predicted overall positive aging after age,

S. Liu, *Social Support Networks, Coping and Positive Aging Among the Community-Dwelling Elderly in Hong Kong*,
https://doi.org/10.1007/978-981-13-3783-3_7

Table 7.1 Hypothesis 1.1–1.5

Hypothesis 1.1: The size of family network has a positive effect on positive aging
Hypothesis 1.2: The size of peer network has a positive effect on positive aging
Hypothesis 1.3: The social network size has a positive effect on positive aging
Hypothesis 1.4: Functional family support has a positive effect on positive aging
Hypothesis 1.5: Functional peer support has a positive effect on positive aging

gender, education and perceived SES were controlled for. The three predictors (e.g., family network size, peer network size and social network size) explained a significant proportion (11.9%) of additional variance in overall positive aging. In terms of the effect of independent variables on five components of positive aging, family network size was positively predictive of CE ($\beta = 0.361, p < 0.001$), PE ($\beta = 0.180$, $p < 0.001$) and life satisfaction ($\beta = 0.130, p < 0.05$). For peer network size, it had a positive effect on each of the five components, that is, illness avoidance ($\beta = 0.190, p < 0.001$), functioning ($\beta = 0.134, p < 0.01$), CE ($\beta = 0.104, p < 0.05$), PE ($\beta = 0.121$, $p < 0.05$), and life satisfaction ($\beta = 0.124, p < 0.05$). Furthermore, the social network size was only positively predictive of functioning ($\beta = 0.123, p < 0.05$) and PE ($\beta = 0.138, p < 0.01$). It could be noted that family/peer/social network size all positively predictive of overall positive aging. However, family and social network size did not significantly affect each component of positive aging. Both family and social network size did not predict illness avoidance. Nevertheless, peer support network size had a positive effect on all the components of positive aging, revealing that a larger peer network would predict better achievement of positive aging. Drawing on the significant predicting effect of family/peer/social network size on overall positive aging, we could say Hypotheses 1.1 and 1.3 were partially supported, Hypothesis 1.2 were fully supported.

The findings of testing of Hypothesis 1.4–1.5 were displayed in Table 7.2. In regard to the regression model of overall positive aging, both functional family and peer support were positively predictive of overall positive aging with a significant variance of 7% explained by the combined dependent variables. Functional family support had a positive effect on overall positive aging ($\beta = 0.121, p < 0.05$), CE ($\beta = 0.309$, $p < 0.001$), and PE ($\beta = 0.122, p < 0.05$). The findings in terms of functional family support revealed that more actual support provided by primary family members were positively associated with better outcome of positive aging, especially with caring and productive engagement. In terms of functional peer support, statistics showed that it had a positive effect on overall positive aging ($\beta = 0.207, p < 0.001$), functioning ($\beta = 0.154, p < 0.01$), PE ($\beta = 0.160, p < 0.01$), and life satisfaction ($\beta = 0.145, p < 0.01$), reflecting that the receipt of more functional support provided peer groups would increase the outcomes of overall positive aging, especially of functioning, PE and life satisfaction. Based on the above findings, Hypothesis 1.4 and 1.5 were partially supported (Table 7.3).

Table 7.2 Results of hierarchical linear regression analysis of hypothesis 1.1–1.3

Independent variables	Dependent variables					
	PA $\beta(SE)$	IA $\beta(SE)$	F $\beta(SE)$	CE $\beta(SE)$	PE $\beta(SE)$	LS $\beta(SE)$
Family network size	0.169*** (0.005)	−0.068 (0.007)	0.040 (0.007)	0.361*** (0.010)	0.180*** (0.010)	0.130* (0.006)
Peer network size	0.190*** (0.005)	0.134** (0.007)	0.150** (0.007)	0.104* (0.010)	0.121* (0.010)	0.124* (0.006)
Social network size	0.133** (0.003)	0.075 (0.005)	0.123* (0.004)	0.022 (0.006)	0.138** (0.007)	0.078 (0.004)
R^2 change	0.119***	0.016*	0.020*	0.132***	0.048***	0.034***

Note PA = overall positive aging, IA = illness avoidance, F = functioning, CE = caring engagement, PE = productive engagement, LS = life satisfaction. Coefficients shown were standardized Age, gender, education and perceived SES were controlled for $*p < 0.05$, $**p < 0.01$, $***p < 0.001$

Table 7.3 Results of hierarchical linear regression analysis of hypothesis 1.4–1.5

Independent variables	Dependent variables					
	PA $\beta(SE)$	IA $\beta(SE)$	F $\beta(SE)$	CE $\beta(SE)$	PE $\beta(SE)$	LS $\beta(SE)$
Functional family support	0.121* (0.029)	−0.065 (0.044)	−0.025 (0.041)	0.309*** (0.061)	0.122* (0.063)	0.100 (0.037)
Functional peer support	0.207*** (0.038)	0.122* (0.057)	0.154** (0.053)	0.071 (0.080)	0.160** (0.082)	0.145** (0.048)
R^2 change	0.070***	0.012	0.020*	0.104***	0.050***	0.038***

Note PA = overall positive aging, IA = illness avoidance, F = functioning, CE = caring engagement, PE = productive engagement, LS = life satisfaction. Coefficients shown were standardized Age, gender, education and perceived SES were controlled for $*p < 0.05$, $**p < 0.01$, $***p < 0.001$

Hypothesis 1.6–1.10 were postulated based on the implications from SST and the empirical findings (Fung, Stoeber, Yeung, & Lang 2008). Table 7.4 is a summary of Hypothesis 1.6–1.10.

To test Hypothesis 1.6–1.7 which concerned age and social network, Pearson correlation analysis was conducted. Results showed that age was positively correlated with size of close social network ($r = 0.182, p = 0.000$) and was negatively correlated with size of peripheral social network ($r = -0.139$, $p = 0.006$). However, the correlation between age and the social network size was non-significant ($r = -0.042$, $p = 0.403$). The evidence reflected that people at an older age tended to have more close social partners and less peripheral social partners, which was consistent with the argument derived from SST. However, no significant association was detected between age and the social network size. Therefore, Hypothesis 1.6 was supported

Table 7.4 Hypothesis 1.6–1.10

Hypothesis 1.6: Age is correlated with size of close social network and peripheral social network, respectively
Hypothesis 1.7: Age is negatively associated with social network size
Hypothesis 1.8: The size of peripheral social network is smaller than that of close social network
Hypothesis 1.9: Older adults receive more emotional support than that of ten years ago
Hypothesis 1.10: Older adults receive more companionship support than that of ten years ago

Table 7.5 One-way ANOVA of the mean of different types of network size among three age groups

Variable	F value (*p*)	(df)	65–74 years Mean (SD)	75–84 years Mean (SD)	85–96 years Mean (SD)
Family network size	0.351	(2, 388)	18.38 (6.85)	17.96 (6.21)	17.63 (5.61)
Peer network size	4.03*	(2, 382)	19.25 (5.93)	18.65 (6.10)	16.64 (6.13)
Close network size	6.669**	(2, 390)	2.99 (2.80)	3.75 (3.80)	5.12 (6.11)
Peripheral network size	4.137*	(2, 390)	7.67 (9.18)	6.32 (7.43)	4.27 (5.54)

Note $^{*}p \leq 0.05$, $^{**}p \leq 0.001$

and Hypothesis 1.7 was not supported. To further investigate the relationship between age and size of different types of social networks, One-Way ANOVA was performed to explore family/peer network size, close/peripheral network size in terms of three age groups, namely, the young-old (65–74 years), the middle-old (75–84 years), and the old-old (85–96 years). Results showed that significant differences existed among three age groups in terms of peer network size (F (382) = 4.03, $p < 0.05$), close network size (F (390) = 6.669, $p < 0.01$), and peripheral network size (F (390) = 4.137, $p < 0.05$). However, there was no significant difference detected between three age groups in terms of family network size (Table 7.5).

In terms of Hypothesis 1.8–1.10, T-test was carried out to examine the problems. To test Hypothesis 1.8, a Paired sample T-test was conducted. Results showed that the mean (M = 3.64) of the size of close social network was significantly smaller than that (M = 6.57) of the size of peripheral social network, $t = -6.846$, $p = 0.000$. Furthermore, One-Sample T-test was conducted to test Hypothesis 1.9 and 1.10. Results showed that the emotional support change (*Mean Difference* = 0.108, t = 3.502, $p = 0.001$) and companionship support change (*Mean Difference* = 0.166, t = 5.194, $p = 0.000$) were significant. This evidence was consistent with the implication of SST stating that older people tended to demand and receive more emotional and companionship support than younger times because of their emotional needs. In summary, Hypothesis 1.8–1.10 were all supported.

Table 7.6 showed Hypothesis 1.11–1.13 which concerned the effect of emotional closeness on positive aging.

Table 7.6 Hypothesis 1.11–1.13

Hypothesis 1.11: Family emotional closeness has a positive effect on overall positive aging
Hypothesis 1.12: Peer emotional closeness has a positive effect on overall positive aging
Hypothesis 1.13: The size of close social network has a positive effect on positive aging

Table 7.7 Hierarchical linear regression analysis of emotional closeness on positive aging

Predictors	Dependent variable (PA)		
	Model 1 (β, *SE*)	Model 2 (β, *SE*)	Model 3 (β, *SE*)
Age	−0.247***(0.004)	−0.218*** (0.004)	−0.213*** (0.004)
F	0.044 (0.067)	−0.049 (0.064)	−0.048 (0.064)
Education	0.089 (0.024)	0.079 (0.023)	0.083 (0.023)
Perceived SES	0.290*** (0.032)	0.242*** (0.031)	0.246*** (0.031)
Family emotional closeness		0.213*** (0.020)	
Peer emotional closeness		0.103* (0.24)	
Integrated emotional closeness			0.259*** (0.025)
R^2(R^2 change)	0.176***	0.242*** (0.066***)	0.240*** (0.064***)

Note F (Female) = 2, M (Male) = 1, PA = overall positive aging
*$p < 0.05$, **$p < 0.01$, ***$p < 0.001$

Hierarchical regression analysis was conducted to test Hypothesis 1.11 and 1.12 (see Table 7.7). Derived from the findings of regression analysis in Model 2, respondents with stronger family emotional closeness (β = 0.213, $p < 0.001$) and peer emotional closeness (β = 0.103, $p < 0.05$) reported better achievement of positive aging after age, gender, education and perceived SES was controlled for. The two factors could explain a significant proportion of 6.6% of the additional variance in positive aging (R^2 *change* = 0.066, $p < 0.001$). Meanwhile, integrated emotional closeness was also found to have a positive effect (β = 0.259, $p < 0.001$) on positive aging. Therefore, Hypothesis 1.11 and 1.12 were supported.

Another hierarchical linear regression analysis was employed to explore the effect of the size of close social network (see Table 7.8). In Model 2, the size of close social network was positively predictive of positive aging (β = 0.142, $p < 0.01$) after age, gender, education and perceived SES were controlled for. The size of close social network could explain 1.9% of the additional variance in positive aging. Furthermore, the effect of the size of close social network was tested in hierarchical linear regression analysis after age, gender, education, perceived SES and social network size were controlled for. Surprisingly, in model 4, the predicting effect of the size of close social network (β = 0.027, p = 0.622) disappeared after inserting the variable *social network size* into the first step. This finding revealed that the size of close social network could positively predict positive aging singly, but its effect disappeared after controlling

Table 7.8 Hierarchical linear regression analysis of the size of close social network on positive aging

Predictors	Dependent variable (PA)			
	Model 1 $\beta(SE)$	Model 2 $\beta(SE)$	Model 3 β (SE)	Model 4 β (SE)
Age	−0.257*** (0.004)	−0.275*** (0.004)	−0.234*** (0.004)	−0.239*** (0.004)
F	−0.051 (0.067)	−0.068 (0.066)	−0.068(0.065)	−0.070 (0.065)
Education	0.085 (.024)	0.088 (0.024)	0.103 (0.023)	0.102 (0.023)
Perceived SES	0.286*** (0.032)	0.265*** (0.032)	0.251*** (0.032)	0.249*** (0.032)
Social network size			0.229*** (0.003)	0.215*** (0.003)
Size of close social network		0.142** (0.008)		0.027 (0.009)
R^2(R^2 change)	0.178***	0.197** (0.019**)	0.229***	0.229*** (0.001)

Note F (Female) = 2, M (Male) = 1, PA = overall positive aging
*$p < 0.05$, **$p < 0.01$, ***$p < 0.001$

for social network size. Then it could be elaborated that the effect of social network size in contributing to positive aging was more prevalent than that of the size of close social network. Hence, we concluded that Hypothesis 1.13 was partially supported.

7.1.2 Testing of Hypothesis Set 2

To test Hypothesis 2.1–2.6 (listed in Table 7.9), Pearson correlation and hierarchical linear regression analysis were conducted relatively.

Results of Pearson correlation analysis showed that age was negatively correlated with problem coping ($r = -0.258$, $p < 0.001$) and emotion coping ($r = -0.191$, $p < 0.01$), respectively. However, no significant correlation was detected between age and avoidance coping ($r = -0.000$, $p > 0.05$). Therefore, Hypothesis 2.1 and 2.2 were not supported. Hypothesis 2.3 was supported.

To further investigate the relationship between age and coping, a One-Way ANOVA was carried out to explore coping in terms of three age groups, namely,

Table 7.9 Hypothesis 2.1–2.6

Hypothesis 2.1: Age is positively correlated with emotion coping
Hypothesis 2.2: Age is positively correlated with avoidance coping
Hypothesis 2.3: Age is negatively correlated with problem coping
Hypothesis 2.4: Problem coping has a positive effect on positive aging
Hypothesis 2.5: Emotion coping has a positive effect on positive aging
Hypothesis 2.6: Avoidance coping has a positive effect on positive aging

Table 7.10 One-way ANOVA of the mean of problem, emotional and avoidance coping among three age groups

Variable	F value (p)	(df)	65–74 years Mean (SD)	75–84 years Mean (SD)	85–96 years Mean (SD)
Problem	12.741**	(2, 380)	0.87 (0.49)	0.71 (0.46)	0.52 (0.44)
Emotion	6.544*	(2, 382)	0.74 (0.43)	0.61 (0.38)	0.55 (0.38)
Avoidance	0.165	(2, 382)	0.79 (0.40)	0.81 (0.38)	0.77 (0.45)

Note *$p \leq 0.01$, **$p \leq 0.001$

the young-old (65–74 years), the middle-old (75–84 years), and the old-old (85–96 years). Findings were shown in Table 7.10. Results showed that significant differences existed among three age groups in terms of problem coping (F (380) = 12.741, $p < 0.001$) and emotion coping (F (382) = 6.544, $p < 0.01$). However, no significant differences in regard to avoidance coping were found (F (382) = 0.165, $p > 0.05$). Hence, this finding revealed that the avoidance coping strategies were evenly adopted among the participants in this sample.

Hierarchical linear regression analysis was conducted to test Hypothesis 2.4–2.6. First, three models containing each type of coping were investigated respectively after age, gender, education and perceived SES were controlled for. In each separated model, statistics showed that problem coping was positively predictive of positive aging ($\beta = 0.289$, $p < 0.001$), so was emotion coping ($\beta = 0.238$, $p < 0.001$) and avoidance coping ($\beta = 0.141$, $p < 0.01$). Consequently, Hypothesis 2.4–2.6 were supported. Furthermore, to further test the combined effect of three types of coping strategies, model 4 presented the results of hierarchical linear regression analysis. The predictive effect of avoidance coping on positive aging became non-significant when all the three types of coping were combined in step 2 (Table 7.11).

Hypothesis 2.7–2.12 were summarized in Table 7.12. This series of hypotheses concerned the mediating effect of coping on the relationship between social support network and positive aging.

To test each of the hypotheses in this section, a mediation analysis was conducted respectively. It appears that a mediational effect occurs when a predictor conceptually causes the mediator and in turn causes the ultimate outcome (Mackinnon, 2000). To carry out a mediation analysis, presumably, there should be significant association(s) between the predictor(s) and outcome variable(s) (Holmbeck, 2002; Mackinnon, 2008). It was hypothesized that the properties of social support network (e.g., size/function of family/peer networks and family/peer emotional closeness) would trigger a specific type of coping strategy, which would contribute to positive aging. Based on this postulation, the mediation model of properties of social support network, coping strategies and positive aging were illustrated (see Fig. 7.1). The mediation tests were conducted in SPSS AMOS 20.0.

Hypothesis 2.7 was tested separately. In terms of the mediating mechanism between the size of family network, problem coping and positive aging, the direct effect of size of family network on positive aging (c'-path, $\beta = 0.24$, $p < 0.001$), and

Table 7.11 Hierarchical linear regression analysis of problem/emotion/avoidance coping on positive aging

Predictors	Dependent variable (PA)			
	Model 1 $\beta(SE)$	Model 2 β (SE)	Model 3 $\beta(SE)$	Model 4 $\beta(SE)$
Age	−0.214*** (0.004)	−0.227*** (0.004)	−0.274*** (0.004)	−0.195*** (0.004)
F	−0.043 (0.065)	−0.042 (0.065)	−0.055 (0.066)	−0.049 (0.065)
Education	0.018 (0.024)	0.060 (0.024)	0.054 (0.024)	0.005 (0.024)
Perceived SES	0.266*** (0.031)	0.265*** (0.032)	0.292*** (0.032)	0.254*** (0.032)
Problem coping	0.289*** (0.060)			0.232*** (0.081)
Emotion coping		0.238*** (0.069)		0.132* (0.086)
Avoidance coping			0.141** (0.071)	−0.012 (0.083)
R^2	0.183***	0.183***.	0.176***	0.176***
R^2 change	0.072***	0.053***	0.020**	0.090***

Note F (Female) = 2, M (Male) = 1. PA = overall positive aging
*$p < 0.05$, **$p < 0.01$, ***$p < 0.001$

Table 7.12 Hypotheses 2.7–2.11

Hypothesis 2.7: Problem coping can mediate the effect of the size/function of family support network on positive aging
Hypothesis 2.8: Emotion coping can mediate the effect of the size/function of family support network on positive aging
Hypothesis 2.9: Emotion coping can mediate the effect of the size/function of peer support network on positive aging
Hypothesis 2.10: Problem coping can mediate the effect of family/peer emotional closeness on positive aging
Hypothesis 2.11: Emotion coping can mediate the effect of family/peer emotional closeness on positive aging
Hypothesis 2.12: Avoidance coping can mediate the effect of family/peer emotional closeness on positive aging

the size of family network was positively associated with problem coping (a-path, $\beta = 0.27$, $p < 0.001$). In addition, the mediator (problem coping) was positively associated with positive aging (b-path, $\beta = 0.31$, $p < 0.001$). Since the three coefficients were significant, the mediating test confirmed the mediating role of problem coping in the relationship between the size of family network and positive aging ($\beta = 0.08$). Furthermore, a Sobel Test was conducted to test the significance of indirect effect ($\beta = 0.08$). Result showed that the indirect effect of the size of family network on positive aging via problem coping was significant ($Z = 3.99$, $p < 0.001$). Therefore, the total effect of the size of family network on positive aging with the mediator was significant (c-path = 0.32, $p < 0.001$), suggesting a partial mediation (see Fig. 7.2). Similarly, the

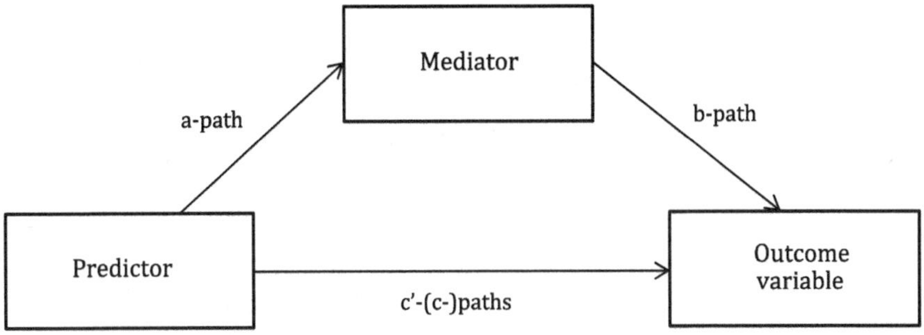

Fig. 7.1 Mediation model of social support network, coping and positive aging. Note: a-path, b-path, c'-path and c-path are all standardized β-coefficients

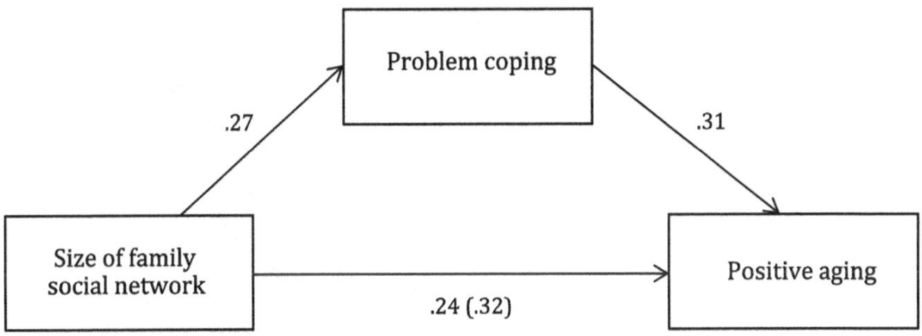

Fig. 7.2 Mediation model of the size of family social network, problem coping and positive aging. Note: All the coefficients were significant standardized values

mediating role of problem coping in the relationship between the function of family network and positive aging was confirmed (a-path = 0.21, $p < 0.001$; b-path = 0.34, $p < 0.001$; c' = 0.19, $p < 0.001$). Sobel test showed that the indirect effect was also significant ($Z = 3.58$, $p < 0.001$). Consequently, Hypothesis 2.7 was supported.

Similarly, mediation tests were conducted to test Hypothesis 2.8 and 2.9. The result showed that the mediating role of emotion coping in the relationship between the size of family network and positive aging was confirmed (a-path = 0.32, $p < 0.001$; b-path = 0.24, $p < 0.001$; c' = 0.25, $p < 0.001$). Sobel test showed that the indirect effect was also significant ($Z = 4.01$, $p < 0.001$). In the same way, the partial mediation model of the function of family network, emotion coping and positive aging was confirmed. Consequently, Hypothesis 2.8 was supported.

The findings of the mediation test of Hypothesis 2.9 confirmed the partial mediation model of the size/function of peer support network, emotion coping and positive aging. Hypothesis 2.9 was supported.

In terms of Hypothesis 2.10–2.12, similar mediation test were carried out. The findings of testing Hypothesis 2.10 and 2.11 confirmed the partial mediation model of family/peer emotional closeness, problem/emotion coping and positive aging. Therefore, Hypothesis 2.10 and 2.11 were supported. However, in terms of Hypoth-

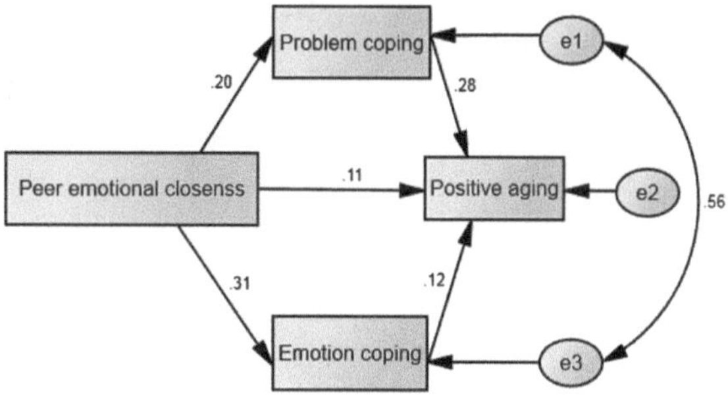

Fig. 7.3 Joint mediation model of peer emotional closeness, problem coping, emotion coping and positive aging. Note: All the coefficients were significant standardized values

esis 2.12, the effect of family emotional closeness on avoidance coping was not significant (a-path = 0.041, $p = 0.411$), failing to verify the mediation model. Similarly, the effect of peer emotional closeness on avoidance coping was not significant, (a-path = 0.074, $p = 0.142$), failing to test the mediation model, neither. It could be found that neither family nor peer emotional closeness could trigger avoidance coping in this sample. Hence, Hypothesis 2.12 was not supported.

Inspired by the testing of Hypothesis 2.10 and 2.11, a joint mediation model was proposed which combined both problem and emotion coping, family and peer emotional closeness and positive aging. The findings supported a joint mediation model of peer emotional closeness, problem and emotion coping, and positive aging (see Fig. 7.3). Sobel Test for the path of peer emotional closeness on positive aging via problem coping confirmed the indirect effect $\beta = 0.09$ ($Z = 3.15$, $p < 0.001$). However, the indirect effect of peer emotional closeness on positive aging via emotion coping was marginally confirmed ($\beta = 0.04$, $Z = 1.929$, $p = 0.054$). Thus, in light of the statistics, we could generally accept the joint mediation model and confirm the direct and indirect effects of peer emotional closeness on positive aging via problem coping and emotion coping.

7.1.3 Testing of Hypothesis Set 3

A set of hypotheses related to demographic variables were tested in this section. Additionally, variables in terms of the employment of DH were also investigated. Table 7.13 presented a total of 13 hypotheses in Hypothesis set 3.

To test Hypothesis 3.1–3.2, Pearson correlation was conducted. The findings showed that age was negatively correlated with overall positive aging ($r = -0.275$, $p < 0.001$). There was no significant relationship between age and employment of DH (operationalized as whether or not employ a DH) ($r = 0.082$, $p = 0.104$). Therefore, Hypothesis 3.1 was supported and Hypothesis 3.2 was not supported. In terms of Hypothesis 3.3–3.7, results of correlation analysis showed that education was positively associated with problem coping ($r = 0.340$, $p < 0.001$), emotion coping ($r = 0.174$, $p < 0.001$), and overall positive aging ($r = 0.220$, $p < 0.001$). No significant correlations were found between education and social network size ($r = -0.039$, p

Table 7.13 Hypotheses 3.1–3.13

Hypothesis 3.1: Age is negatively correlated with positive aging
Hypothesis 3.2: Age is positively correlated with the employment of domestic helper
Hypothesis 3.3: Education is positively correlated with emotion coping
Hypothesis 3.4: Education is positively correlated with avoidance coping
Hypothesis 3.5: Education is positively correlated with problem coping
Hypothesis 3.6: Education is positively correlated with positive aging
Hypothesis 3.7: Education is positively correlated with social network size
Hypothesis 3.8: Perceived SES is positively correlated with social network size
Hypothesis 3.9: Perceived SES is positively correlated with positive aging
Hypothesis 3.10: Perceived SES is positively correlated with employment of DH
Hypothesis 3.11: Employment of DH is positively correlated with positive aging
Hypothesis 3.12: Among the older adults who employ DHs, satisfaction with domestic helper's care provision is positively correlated with positive aging
Hypothesis 3.13: Older adults who employ DH have less intention to stay at residential care homes, compared with those without DH

= 0.438) and education and avoidance coping ($r = 0.097$, $p = 0.058$). Consequently, Hypothesis 3.3, 3.5 and 3.6 were supported. Hypothesis 3.4 and 3.7 were not supported. The findings of testing of Hypothesis 3.8–3.10 indicated that perceived SES were positively related with social network size ($r = 0.147$, $p < 0.01$), positive aging ($r = 0.283$, $p < 0.001$), and the employment of DH ($r = 0.149$, $p < 0.01$). Hence, Hypothesis 3.8–3.10 were supported.

In terms of Hypothesis 3.11 and 3.12, the findings of correlation analysis showed that employment of DH was negatively related with overall positive aging ($r = -0.126$, $p < 0.05$), indicating that older adults who employed DH tended to report lower score of positive aging. Furthermore, correlation analysis was conducted between employment of DH and the five components of positive aging. The results indicated that employment of DH was negatively correlated with illness avoidance ($r = -0.135$, $p < 0.01$) and functioning ($r = -0.140$, $p < 0.01$). Thus, it could be noted that the employment of DH was only related to the health and functioning facets of positive aging. In terms of Hypothesis 3.12, statistics showed that there was no significant relationship between satisfaction of DH's care provision and positive aging ($r = -0.134$, $p = 0.496$). Consequently, Hypothesis 3.11 and 3.12 were not supported.

To test Hypothesis 3.13, a T-test was employed. The finding indicated that the whether to employ DHs or not had no significant difference in terms of showing willingness to stay at residential care homes (t (388) = 0.081, $p = 0.935$). Therefore, Hypothesis 3.13 was not supported.

Table 7.14 Summarized results of hypothesis testing

No.	Content of hypothesis	Testing result
1.1	The size of family network has a positive effect on positive aging	√×
1.2	The size of peer network has a positive effect on positive aging	√
1.3	The social network size has a positive effect on positive aging	√×
1.4	Functional family support has a positive effect on positive aging	√×
1.5	Functional peer support has a positive effect on positive aging	√×
1.6	Age is correlated with size of close social network and peripheral social network, respectively	√
1.7	Age is negatively correlated with social network size	×
1.8	The size of peripheral social network is smaller than that of close social network	√
1.9	Older adults receive more emotional support than that of ten years ago	√
1.10	Older adults receive more companionship support than that of ten years ago	√
1.11	Family emotional closeness has a positive effect on overall positive aging	√
1.12	Peer emotional closeness has a positive effect on overall positive aging	√
1.13	The size of close social network has a positive effect on overall positive aging	√×
2.1	Age is positively correlated with emotion coping	×
2.2	Age is positively correlated with avoidance coping	×
2.3	Age is negatively correlated with problem coping	√
2.4	Problem coping has a positive effect on positive aging	√
2.5	Emotion coping has a positive effect on positive aging	√
2.6	Avoidance coping has a positive effect on positive aging	√
2.7	Problem coping can mediate the effect of the size/function of family support network on positive aging	√
2.8	Emotion coping can mediate the effect of the size/function of family support network on positive aging	√
2.9	Emotion coping can mediate the effect of the size/function of peer support network on positive aging	√
2.10	Problem coping can mediate the effect of family/peer emotional closeness on positive aging	√
2.11	Emotion coping can mediate the effect of family/peer emotional closeness on positive aging	√
2.12	Avoidance coping can mediate the effect of family/peer emotional closeness on positive aging	×
3.1	Age is negatively correlated with positive aging	√
3.2	Age is positively correlated with the employment of domestic helper	×
3.3	Education is positively correlated with emotion coping	√

(continued)

Table 7.14 (continued)

No.	Content of hypothesis	Testing result
3.4	Education is positively correlated with avoidance coping	×
3.5	Education is positively correlated with problem coping	√
3.6	Education is positively correlated with positive aging	√
3.7	Education is positively correlated with social network size	×
3.8	Perceived SES is positively correlated with social network size	√
3.9	Perceived SES is positively correlated with positive aging	√
3.10	Perceived SES is positively correlated with employment of DH	√
3.11	Employment of DH is positively correlated with positive aging	×
3.12	Among the older adults who employ DHs, satisfaction with domestic helper's care provision is positively correlated with positive aging	×
3.13	Older adults who employ DH have less intention in terms of staying at residential care homes, compared with those without DH	×

Note √ represents the hypothesis was supported. × represents the hypothesis was not supported. √ × means the hypothesis was partially supported

7.2 Summary

In this chapter, a total of 38 hypotheses were tested. The analytical process and statistical findings were presented to provide evidence of hypothesis testing. Finally, Table 7.14 summarized the results of hypothesis testing.

References

Fung, H. H., Stoeber, F. S., Yeung, D. Y. L., & Lang, F. R. (2008). Cultural specificity of socioemotional selectivity: Age differences in social network composition among Germans and Hong Kong Chinese. *The Journals of Gerontology, Series B: Psychological Sciences and Social Sciences, 63*(3), 156–164.

Holmbeck, G. N. (2002). Post-hoc probing of significant moderational and mediational effects in studies of pediatric populations. *Journal of Pediatric Psychology, 27*(1), 87–96.

Mackinnon, D. P. (2000). Contracts in multiple mediator models. In J. S. Rose, L. Chassin, C. C. Presson, & S. J. Sherman (Eds.), *Multivariate applications in substance use research: New methods for new questions* (pp. 141–160). Mahwah, New Jersey: Erlbaum.

Mackinnon, D. P. (2008). *Introduction to statistical mediation analysis*. New York: Taylor & Francis.

Chapter 8
Domestic Helper as a Support Provider

This chapter displays the major findings in relation to the social support provided by domestic helper, the extent of satisfaction towards services provided by domestic helper, the existing relationship with positive aging and the attitudes of employing domestic helper in taking care of the elderly. Then the discussion of the above points is presented.

8.1 Support Provided by Domestic Helpers

8.1.1 Descriptive Statistics of Respondents Who Employ Domestic Helpers

In the present sample, a total of 393 older adults participated in the survey study. Of all the participants, 31 older people reported that they employed domestic helper at the time they were interviewed, accounting for 7.9% of the total sample. Among the elderly who employed domestic helper, 23 respondents employed live-in foreign domestic helper, 7 respondents employed part-time local helper and only 1 participants employed full-time local helper, accounting for 74.2, 22.6 and 3.2% of the group who employed, respectively. As Table 8.1 shows, 22.6% of the respondents who employed domestic helper replied that they did not need caregivers because they had self-care ability. Adult children and spouses as caregivers account for a small percentage (22.5%). And over half of the domestic helpers were reported serving as primary caregivers for the elderly. Findings revealed that the rate of employing domestic helper (7.9%) among the group of community-dwelling older adults was a bit smaller than the rate (12.2%) of older persons who had paid helpers according to the statistics from the Thematic Household Survey Report (2009).

S. Liu, *Social Support Networks, Coping and Positive Aging Among the Community-Dwelling Elderly in Hong Kong*,
https://doi.org/10.1007/978-981-13-3783-3_8

Table 8.1 Primary caregivers of respondents who employed domestic helper (N = 31)

	NIL	Spouse	Son	Daughter	Daughter-in-law	Domestic helper
Frequency	7	1	4	1	1	17
Percent (%)	22.6	3.2	12.9	3.2	3.2	54.8

Note NIL represents respondents without need of caregiving

Table 8.2 Perceived SES of respondents who employed domestic helper (N = 29)

	Lower	Lower-middle	Middle	Upper
Frequency	4	11	13	1
Percent (%)	12.9	35.5	41.9	3.2

Table 8.3 Personal monthly income of respondents who employed domestic helper (N = 31)

	None	Below 2000	2001–4000	4001–6000	10,000 and above	Not clear
Frequency	1	6	13	5	2	4
Percent (%)	3.2	19.4	41.9	16.1	6.5	12.9

Note monthly income is calculated by Hong Kong Dollar

Table 8.4 Living arrangement or cohabitant of respondents who employed domestic helper (N = 31, multiple responses)

	Domestic helper (solely with)	Spouse	Son	Daughter	Daughter-in-law	Son-in-law	Grand-children
Frequency	7	5	6	4	7	2	3
Percent of cases (%)	28.0	20.0	24.0	16.0	28.0	8.0	12.0

Table 8.2 displays the situation of perceived SES of the respondents who employed domestic helpers. It could be drawn that nearly half of the participants are from the lower and lower-middle class. About 41.9% of them were from the middle class.

In Table 8.3, the situation in relation to personal monthly income of the elderly who employed domestic helper is exhibited. In accordance with the perceived SES shown by Table 8.2, most of the elderly who employ domestic helper converge on a monthly income around HKD 2001-4000, representing the income level from the lower-middle class.

Multiple responses towards living arrangement or cohabitant of older respondents who employed domestic helper are presented in Table 8.4. Statistics showed that 28% of the elderly lived with foreign domestic helper solely. Others reported they may live with adult children and grandchildren together except for foreign domestic helper.

Table 8.5 Caregiving and support types provided by domestic helpers (N = 31, multiple responses)

	Basic housework	Personal care	Escort (go out to see a doctor)	Assisting in medical care and nursing	Chatting	Caring for little children and family members
Frequency	31	12	18	8	9	5
Percent of cases (%)	100.0	38.7	58.1	25.8	29.0	16.1

Table 8.6 Payments of wage to domestic helpers (N = 30, multiple responses)

	Oneself	Son	Daughter-in-law	Daughter	Son-in-law
Frequency	6	13	2	16	2
Percent of cases (%)	20.0	43.3	6.7	53.3	6.7

Note Oneself means full payment by the elderly themselves

In terms of social support provided by domestic helper, it could be learned from Table 8.5 that *basic housework* (100% of all cases) and *escort to see a doctor* (58.1% of all cases) were the most prevalent tasks performed by domestic helper, followed by *personal care* (38.7% of all cases), *chatting* (29% of all cases) and *assisting in medical care and nursing* (25.8% of all cases).

In terms of the wage payer, Table 8.6 exhibited the details of payment from the elderly and other family members. It could be found that son (43.3% of all cases) and daughter (53.3% of all cases) afford most of the payment other than the elderly themselves (20% of all cases). Children-in-law only account for a small share of the payment in this sample.

8.1.2 *Extent of Satisfaction Towards Services Provided by Domestic Helpers*

For the question enquiring the degree of satisfaction towards services provided by domestic helpers, a total of 30 valid answers were recorded (see Table 8.7). A number of 14 respondents reported neutral attitude in terms of service satisfaction, accounting for 46.7%. Additionally, 8 and 4 respondents reported satisfactory and extremely satisfactory attitudes towards the domestic helper respectively, accounting for 40% of the total. 2 participants showed extremely not satisfactory and not satisfactory respectively, accounting for 13.3%.

Table 8.7 Extent of satisfaction towards services provided by domestic helpers (N = 30)

	Extremely not satisfied	Not satisfied	Normal	Satisfied	Extremely satisfied
Frequency	2	2	14	8	4
Percent (%)	6.7	6.7	46.7	26.7	13.3

Table 8.8 Employment of domestic helper and group means of positive aging (N = 393)

Variable	t value (*p*)	(df)	With domestic helper	Without domestic helper
IA	2.700**	(1, 390)	3.05	3.47
F	2.773**	(1, 393)	3.08	3.49
CE	0.895	(1, 390)	2.32	2.53
PE	1.294	(1, 390)	2.37	2.67
LS	1.288	(1, 389)	3.33	3.51
E-PAI	2.481*	(1, 379)	2.97	3.25

Note IA = Illness Avoidance, F = Functioning, CE = Caring Engagement, PE = Productive Engagement, LS = Life Satisfaction, E-PAI = expanded Positive Aging Index. *$p < 0.05$; **$p < 0.01$

8.2 Employment of Domestic Helper and Positive Aging

In this section, T-test of means in terms of overall positive aging and its components were conducted to look at the group differences between respondents who employed domestic helper and respondents without domestic helper (Table 8.8).

Results showed that respondents with domestic helper reported significant poorer conditions in terms of IA and Functioning. There is no distinct difference between the two groups in terms of CE, PE and LS. However, with regard to the overall expanded Positive Aging Index, older adults who employed domestic helper reflected lower score than the other group.

8.3 Pros and Cons of Employing Household Domestic Helper

In terms of the willingness to employ domestic helper, the respondents were asked about their intention whether they would employ a domestic helper if their health conditions were getting worse. A total of 342 answers were collected, and 74 of them (18.8%) showed positive attitude to employ a domestic helper, 164 of them (41.7%) indicated no willingness and 104 participants (26.5%) were not clear about this question. To further explore the appraisals of employing a domestic helper and the pros and cons of the services provided by them, participants were asked to express their thinking about the advantages and disadvantages of employing a domestic

Table 8.9 Categorized attitudes towards employment of domestic helper*

Category	No.	Attitudes	Frequency
Financial	1	The wage of domestic helper is high so that they cannot afford	59
	2	They are lacking of money to employ	59
Service quality	3	Domestic helpers could make the living environment better	56
	4	They are not satisfied with the services provided by helper generally	12
	5	Domestic helpers are lazy and not responsible	6
	6	They are not satisfied with the food taste cooked by helper	3
	7	Domestic helpers are indeed helpful.	2
Communication	8	They find it hard to communicate with foreign domestic helper because of language obstacles	28
	9	FDHs have bad temper	5
Relationship	10	FDHs are not trusted (e.g. stealing, not obedient)	19
	11	FDHs are troublesome	8
Others	12	The space is limited to have a live-in domestic helper	4

Note *294 out of 393 participants answered this question. Answer items with high frequency were selected

helper. The question is open ended. Similar answers were classified into categories. Table 8.9 displayed the sorted prevalent attitudes derived from the answers.

Findings of the attitudes reflected that there were obvious advantages and disadvantages of employing domestic helper. However, attitudes mostly emerged into the shortcomings of having a domestic helper. For the benefits of having a domestic helper, 56 participants agreed that domestic helpers were helpful and could make the living environment better. For the negative attitudes towards domestic helper, 59 participants stated the wage paid to domestic helper was too high and another 59 participants felt they could not afford to employ. There were 28 answers pointing out the communication problem with domestic helpers. 19 respondents thought FDHs were not trustworthy. Additionally, a small number of respondents expressed that they were not satisfied with the food that domestic helpers cooked, bad temper or laziness, etc.

8.4 Discussion and Summary

In this section, discussions on domestic helper in providing social support are presented. Implications of various support provision by domestic helper and the relationship between domestic helper employment and positive aging is discussed based on the above findings. Finally, future directions with regard to the large pool of employing foreign domestic helper in elder care are elaborated.

8.4.1 A Variety of Support Provided by Domestic Helper

It could be noted from the findings that domestic helper including foreign domestic helper and local maids are able to provide various and necessary support or services for the older adults. A majority of the elderly have live-in FDHs, from whom they obtain instrumental support (e.g., basic housework, escorting, personal care and assisting in medical care and nursing etc.). Additionally, in line with the answers collected from open questions, a portion of the older people who employ domestic helper were almost content with the tangible support provided by them. Foreign domestic helper who lives with other family members other than the elder person can also shoulder some caring work for little children or others. Older adults are aware of the important role of domestic helper in supporting them or the household because of their involvement in various caring and household work.

Specifically, the role of domestic helper in caring older adults is essential. By providing services to the elderly in need, the involvement of domestic helper is beneficial for maintaining the elderly in the community. Additionally, they provide much companionship to the elderly when they go out, accompanying and warding them in many circumstances. In the Hong Kong context, adults are usually under different distress from their career and family so that they are unable to provide around-the-clock care to their frail older parents. What's more, the long queue of waiting to enter residential institutions also calls for older adults to stay in the community. The involvement of domestic helper especially forging domestic helper who works as full time maids can provide much help in caregiving for older adults and alleviating the caregiving burden of primary family members.

However, the questions or drawbacks of domestic helpers in caregiving should not be neglected. First, from the views of older adults, most of them are ready to employ maids if they do not need to consider financial issues. Hence, the promotion of recruiting foreign domestic helper in the future might be largely dependent on a reasonable wage. Second, many complaints from the older people converge on the language and communication problems. It might be difficult for the older adults who cannot speak English to communicate freely with the foreign maids. Apart from that, cultural diversities between two regions would constitute an impassable obstacle in bridging the relationship between both parties. The elderly would find it hard to adjust to the taste of food cooked by foreign domestic helper, to build on trust with them and to treat them like family members. Thus, the shortcomings of recruiting FDHs constitute a challenge in promoting the recruitment of foreign domestic helper.

8.4.2 The Relationship Between Employment of Domestic Helper and Positive Aging

In comparison with the elderly who did not employ domestic helper, older adults recruiting domestic helper reported poorer physical and functioning health as well as the overall positive aging. Although the involvement of domestic helper could assist in household work and caregiving, the health condition of older adults could not be changed, which still serves as a hamper to develop caring engagement and productive engagement and to achieve positive aging.

8.4.3 A Large Pool of Employing Domestic Helper in Elder Care in the Future

Though recruiting a domestic helper in this study was found not to facilitate positive aging of the older adults, the role of domestic helper cannot be ignored. It is accepted as a reasonable approach to promote the elderly especially frail older people to age in place, giving more opportunities for the elderly to involve in the setting of community and society. In the long run, importing domestic helper could make contributions to the implementation of long term care for the elderly in Hong Kong by alleviating the pressure from the government, community and primary family members (spouse, adult children, etc.). Hence, there is a large pool of potential to employ domestic helper in elder care in the future.

Recommendations and propositions in relation to promoting domestic helper employment may go to the below points. First of all, it is suggested for the government and relevant sectors to negotiate the agenda of bringing in FDHs in the field of elder care. It is suggested to reconsider the feasibility of providing subsidy to employing domestic helper in elder care so that more eligible families could afford the expense. Secondly, specific and professional training programs are recommended to be organized by social welfare sectors to facilitate the knowledge and skills of foreign maids in taking care of the elderly. A solid background of caregiving skills is very important if the elderly are in need of long term care. The domestic helpers who are equipped could better assist professionals (e.g., doctors, nurses, physicians) in caring the frail elderly. Additionally, domestic helpers are suggested to learn more skills in cooking Chinese food, which could cater for the taste of Chinese families.

Last but not least, measures are suggested to facilitate the interrelationships between the elderly and domestic helper. A recent study of Hong Kong (Chong et al. 2014) has found that a harmonious working relationship between two parties can enhance the service quality and service satisfaction. Therefore, it is suggested that efforts should be made to enhance elderly-helper relationship. The elderly and their primary family members can attempt to treat domestic helpers as proxy family members and maintain a steady connection with them. Appropriate emotional closeness between helpers and the whole family is suggested to cultivate so that better service provision and appraisal could be reached. For the care recipients and their family members, they are encouraged to make regular contact with the helpers, to show concern to them, and to present small gifts to them in order to enhance maids' feeling of being accepted.

8.4.4 *Summary*

This part discussed the implications from the findings of domestic helper as a support provider to the elderly, elaborated the contributions made by domestic helper and proposed the feasible measures and efforts to facilitate long-term care scheme in the future. Endeavors from the policy makers, social welfare sector and the elderly-helper relationship poles were recommended to achieve the goal of aging in place.

References

Chong, A. M. L., Kwan, C. W., Chi, I., Lou, V. W. Q., & Leung, A. Y. M. (2014). Domestic helpers as moderators of spousal caregiver distress. *Journals of Gerontology, Series B: Psychological Sciences and Social Sciences*, *69*(6), 966–972.

Thematic Household Survey Report No. 40. (2009, June 4). *Socio-demographic profile, health status and self-care capability of older persons, census and statistics department*. Retrieved online at http://www.statistics.gov.hk/pub/B11302402009XXXXB0100.pdf.

Chapter 9
Discussion

9.1 Overview

This study has examined the attributes of social support networks, coping strategies, and the expanded Positive Aging Index (PAI). This chapter has two foci. The first focus is to identify and interpret the main findings of the study. The second is to link the observations and interpretations of the sample data with existing theories and past empirical studies.

This chapter is composed of five parts: (1) Expanded Positive Aging Index (PAI) as a validated tool to measure well-being in later life, (2) the interpretation of social network compositions and the prevalence of peer support, (3) the effect of social support networks on positive aging, (4) the significant functions of social support, and finally (5) coping in later life and its mediating effects on social support networks and positive aging.

9.2 Expanded Positive Aging Index (PAI) as a Validated Tool to Measure Well-Being in Later Life

Using a survey of 393 community-dwelling older adults in Hong Kong, the present study adopted an expanded Positive Aging Index (PAI) based on a previously validated tool to measure the main dependent variable—positive aging. The modified 19-item questionnaire is a two-factor model combining the original PAI developed by Ng et al. (2011) with a supplementary component—life satisfaction. It was verified that the 19-item expanded PAI consisted of five factors—illness avoidance, functioning, caring engagement (CE), productive engagement (PE), and life satisfaction. Results of the confirmatory factor analysis showed that this model fit the present data well.

S. Liu, *Social Support Networks, Coping and Positive Aging Among the Community-Dwelling Elderly in Hong Kong*,
https://doi.org/10.1007/978-981-13-3783-3_9

9.2.1 Expanded PAI as a Validated Measurement Tool for Chinese Community-Dwelling Older Adults

In previous studies, positive aging has been researched in both qualitative and quantitative ways as a concept that is similar to successful aging. This occurred because successful aging has been a dominant approach to depict the later life of older adults ever since Rowe and Kahn (1997) first developed the concept. A series of similar conceptions in social gerontology were elaborated with reference to the construct of successful aging. Although debates on the definition and differentiation of successful aging and positive aging have persisted for years, Chinese scholars have made great efforts to explore the concept of positive aging, including analyzing the contents of the concept and validating an appropriate measurement based on Chinese folklore (Chong, Ng, Woo, & Kwan, 2006; Ng et al., 2011). Based on the PAI that was validated in a longitudinal study in Hong Kong, the present study verified the original tool of positive aging after deleting one item that asked, "Have you ever participated in paid work after retirement?" In 2006, focus-group discussions on the domains of positive aging revealed that maintaining active engagement with an activity or with society was an important element. Additionally, productive engagement (i.e., paid or unpaid activities that create goods or services of value) was also addressed in Hong Kong society because of its reflection of Confucian ethics of diligence and social contribution (Le, Shen, & Du, 2004). However, in the present study, the data did not support the factor of participation in paid work with a significant factor loading in the preliminary model test. This finding could be interpreted from the characteristics of the study sample. The present study adopted a non-probability sample that is composed of members in an elderly community center. The respondents who actively participated in this survey mostly keep in touch with the center and regularly attend activities or meetings organized by the center. Therefore, it is supposed that the respondents generally have little engagement in paid work because they have already devoted much energy and time to life in the community center. Furthermore, the average age of the sample participants was 77.3 years old. People of this age tend to have less engagement in paid work because of the job-related age restrictions.

To conclude, the elimination of the item of paid work in this study is mainly attributed to the nature of the study sample. Although this fact could not be theoretically explained, the scale after removing that factor achieved a better model-fit when adjusted to the present sample. In addition, this implication may result in a revisitation of the structure of CE under positive aging among older Hong Kong adults. Given the bias of the present sample, further verification of this tool is needed using a random sample to balance the interior diversities of the sample in future studies.

9.2.2 Expanded PAI as a Tool to Measure a Two-Factor Model of Positive Aging

In terms of the newly added subjective part of the expanded PAI, the subscale of life satisfaction was also validated as a component in the construct of positive aging in this study. Dating back to the theoretical development of successful aging, subjective appraisals of an individual's life have been recorded (Fisher, 1995; Rowe & Kahn, 1987; Ryff, 1982). Most views usually converge on the three-factor construct of successful aging by Rowe and Kahn (1997)—the ability to maintain low risk of disease and disease-related disability, high levels of mental and physical health, and active engagement with life. Some literature focuses on whether it is appropriate to adopt objective or subjective assessments to measure successful aging. In this study, life satisfaction was accepted as a new and important subjective criterion, together with the other components of PAI, to form the expanded PAI. Second-order CFA results lent support to this construct using the present sample. Consequently, the expanded PAI broadens the concept of positive (successful) aging in Hong Kong and depicts a more comprehensive construct involving both objective and subjective criteria. It is in line with the evidence found by Pruchno et al. (2010), who proposed and tested a conceptual two-factor model of successful aging in the US that includes objective and subjective assessments.

For a long time, researchers have endeavored to elucidate the meaning of "success" in regard to the concept successful aging and proposed countless measures of it. A burgeoning number of studies in the past decades defined success as survival, lack of disability, life satisfaction, social engagement, productivity, quality of life, and the absence of disease (Bowling, 2007; Depp & Jeste, 2006). However, few measures combining the comprehensive domains of successful aging were put forward. Although positive aging and successful aging are quite similar and sometimes regarded as equivalent in social gerontology, I chose to explore positive aging because the term "positive" is embedded with slightly different implications.

In my construct, positive aging encompasses more than the avoidance of physical and functional deterioration, and is permeated with positive attitudes and appraisals of later life. In this regard, life satisfaction would be considered as a reasonable criterion to depict positive aging. In the two-factor model of Pruchno et al. (2010), successful aging encompassed both the objective and subjective components. The objective components of successful aging include having few chronic diseases, ample functional ability, and little or no pain. The subjective components include how well they are aging, how successful their aging experience is, and the extent to which they rate their current life as positive. In terms of the present study, caring engagement, productive engagement, and life satisfaction are supposed to meet the above criteria as subjective components. These elements are regarded as an evaluation that individuals make of their own aging experience at one point in time (Pruchno et al. 2010). Thus, the adopted two-factor model in this study is appropriate for the examination of positive aging among older adults. On the one hand, the validated, expanded PAI adds rigor to the measurement of a construct that has intrigued gerontologists for hundreds of years by providing the empirical evidence on which to build research about successful aging; on the other hand, it enriches the implications of positive aging by highlighting the importance of subjective appraisals towards later life.

9.3 The Interpretation of Social Network Composition and the Prevalence of Peer Support

This section mainly describes the important findings in terms of social network compositions and sheds light on the corresponding explanations. Additionally, it will elucidate the implications in terms of the prevalence of peer support that is extracted from the findings of the in-depth interview and the main survey.

9.3.1 The Interpretation of Social Network Composition in the Survey

This study used the Social Convoy Questionnaire (Kahn & Antonucci, 1980) to examine the structure of social support networks among the community-dwelling elderly in Hong Kong. Through this questionnaire, the results are displayed in terms of the total social network size, the network size of the inner, middle and outer social circles, and the number of close and peripheral social partners. Results of the social network analysis showed that the network size of the inner, middle and outer circles reported by the participants were 3.6, 3.0 and 3.6, respectively.

When exploring social relationships within an individual's social network, it is a common practice of scholars to describe the network size as falling into specific groups or characteristics, and then to attempt to interpret the pros and cons of the network mode (Antonucci, 2001). Bosworth and Schaie (1997) pointed out that the structure of support patterns may be particularly useful for gerontological research because of the increasing differentiation among the elderly, and because of its advantage in allowing researchers to examine the heterogeneity of their samples. In retrospect, quite a few researchers have examined the social network compositions among Chinese older adults (Cheng, Lee, Chan, Leung, & Lee, 2009; Fung, Stoeber, Yeung, & Lang, 2008). For the present study, the total network size (10.2) in this study was between the family-focused network type (13.2) and the distant-family network type (8.9), as reported in a study of Hong Kong older adults (Cheng et al., 2009). In 2009, Cheng and associates discovered five social network types among Hong Kong older adults using a K-means cluster analysis method. The five types are diverse, friend-focused, family-focused, restricted, and distant-family. A family-focused network was defined as frequent contact with immediate kin and a high degree of support exchange with them. A distant-family network type was defined as having high levels of contact and support exchange with distant kin, average levels with nonfamily, and comparatively sparse contact and support interaction with immediate kin. The present study indicates an intermediate pattern of social networks in this sample in which interrelationships of family members (primary and extended family members) are valued. It shows that participants in this study tend to have reliable interrelationships with primary family members and maintain normal contacts with extended family members, suggesting that family members constitute an important part of the social network. This finding is consistent with empirical evidence that family support is the primary source of support for the Asian elderly because of the central value of filial piety (Cheng et al., 2009; Deng, Hu, Wu, Dong, & Wu, 2010). Nevertheless, it should be noticed that this inference is probably not sufficiently rigorous because only one feature, namely network size, cannot represent a specific network pattern. More research aiming to discover the characteristics of social network types is encouraged to uncover further details.

As to the evidence in terms of the distribution of social network members in the three circles, one can see the participants have very similar network sizes in terms of the inner and outer circles, which reveals that older adults also have peripheral social relationships. The number of close social partners and peripheral social partners in this sample were 3.64 and 6.57, respectively. This result reflects a smaller size of social networks in comparison to an empirical study in Hong Kong. In Fung et al.'s (2008) study aiming to unravel the characteristics of social network composition among Hong Kong citizens, the findings showed that the respondents in that sample, aged between 20 and 91 years, had 5.11 persons in close social networks and 8.12 in peripheral social networks. It could be inferred that the general number of close and peripheral social partners among the older group is smaller than that among a mixed group of both younger and older adults. As to the age effect on the size of different social networks, the study found that significant differences existed between the three age groups, namely, the young-old (65–74 years), the middle-old (75–84 years), and the old-old (85–96 years). Participants at an older age tended to have smaller peer network size and peripheral network size, but a larger close network size. These findings share some similarities with the proposition derived from socioemotional selective theory (Carstensen, Isaacowitz, & Charles, 1999), which claims that older adults tend to have fewer social partners because of the decreasing demand of career development and social experience. However, because of sampling constraints, the findings of social network composition in this study cannot identify the trajectory of social network changes of older adults compared to their younger counterparts. To address this research gap, longitudinal studies are recommended to verify the hypotheses of lifespan theories in social gerontology, especially the changes of network composition within one's life course.

9.3.2 The Prevalence of Peer Support Among Community-Dwelling Older Adults

The findings of the survey showed that primary family and peer group members accounted for 54.6 and 33.0%, respectively, of the total network. The proportion of primary family members in this study is similar to the result (43.6%) obtained in a previous sample of Hong Kong citizens (Fung et al., 2008). It reveals that family members are the major source of social networks among the elderly. Additionally, the peer group proportion (33%) is remarkably higher than that found in a study in nursing home elderly (11.3%). Given the fact that the community-dwelling elderly have more opportunities to form social networks with peer group members in comparison to the elderly living in nursing homes, this finding is reasonable, although it somewhat contradicts the original postulate of socioemotional selectivity theory (Carstensen et al., 1999), which states that people in later life have a decreased network of peripheral social partners because they may regard their future time as limited. On the other hand, the finding suggests a new insight into interpreting the social networks of older people. Older people may have a greater desire to develop new, emotionally meaningful social partners when key network members are no longer available (e.g., due to illness, death and geographic distance). The participants in this sample all live in the community and they are members of elderly community centers. Thus, they have more opportunities to get to know partners through community networks and find peer group members as emotionally close friends. Thus, a larger number of outer circle-containing center members, old friends, neighbors, and non-kin members would constitute resources to provide emotional and companionship support (Cheng et al., 2009). This finding is also consistent with previous research showing that friends are less likely to maintain negative interactions in daily life because friendships are not bound by prescribed roles and the reciprocal interactions would bring more positive appraisals (Merz & Huxhold, 2010).

In terms of the in-depth interview before the survey was conducted, the thematic findings related to social network composition also revealed the prevalence of peer support among community-dwelling older adults. Transcripts from the in-depth interviews showed that a majority of older respondents valued the role of peer groups composed of old friends, workmates, center members, and so on. The quotes are as follows. "I live alone and have no children. I still keep in touch with my old colleagues and friends. Sometimes we'll talk about worries and chores. (C-8)"; "My children are all living abroad and seldom come back home. If I have some bad mood or unhappiness, I'll turn to my friends in Hong Kong and they are always ready to listen to me and help me. I'm so grateful for this (C-10)." Similar quotations mentioned the importance of peer groups in the daily life of older adults living in the community, especially those who did not live with adult children. The findings revealed that the receipt of peer support provided companionship to older people, especially the widowed, and contributed to loneliness distraction. Additionally, older people gained emotional support from mutually sharing trivial hassles and illness experiences. Lastly, they obtained lots of entertainment from activities and communications with peer partners. In summary, the qualitative findings from the in-depth interview of the role of peer groups lend much support to previous arguments regarding the four reasons, put forth by Pinquart and Sorensen (2000), that friendship support has a strong relationship to psychological well-being in later life. The first reason is that friendships are voluntary, meaning that an established friendship is more likely to be used for emotional support. In the present study, center membership was built on the basis of the identity of center values and regulations. Consequently, center members voluntarily form a natural peer group, among which reciprocal behaviors and interactions occur. The second reason is that friends are usually of the same age and share similar characteristics. It seems that this reason also applies to the nature of the present sample. A third possible reason is that friends tend to be a source of enjoyment and socializing, which could explain the fact that center members like to share the happiness derived from taking part in amusements. However, little evidence from the in-depth interview could prove the fourth explanation that friendships are almost always of high quality since older people are inclined to disengage from unsatisfactory relationships.

In summary, the in-depth interview findings related to the role of peer groups provide further explanations of the evidence found in the survey mentioned earlier, and they have provided information to help interpret the survey findings.

9.3.3 The Preference of Companionship Support from Peer Groups

Some in-depth interviews reported that older adults obtained happiness by getting along with center members in daily activities. This fact infers the demand and preference of receiving companionship among the elderly. Generally, participants said they did anticipate emotional communications between peer groups, but the need for emotional interaction could be compensated by companionship support (e.g., attending activities together, saying hello to others every day, having tea with others) when the object of the communication was not mature or appropriate. This intriguing finding is basically in line with the implications of socioemotional selectivity theory, that is, older adults tend to acquire more emotional closeness from social partners because they can provide emotional and companionship support (Carstensen et al., 1999; Fung et al., 2008; Lang & Carstensen, 1994). Additionally, this finding is also consistent with the western argument that friend-focused networks tend to be high on emotional support, because friends are generally age-peers (Adams & Blieszner, 1995; Crohan & Antonucci, 1989), but low on instrumental support. Furthermore, it should be noted that the function of peer support is of greater importance when the support recipient is widowed or living alone. Geographical distance from adult children or kin has hampered regular communication to a large extent, consequently, peer groups have become resources of emotional and companionship support. This implication is partially consistent with the argument of the hierarchical-compensatory model (Cantor, 1979). In the hierarchical-compensatory model, an older person is at the center of a series of concentric circles, ranging from informal at the center to formal at the periphery, each containing a different kind of support. Older people interact with each of these circles and sub-systems at different times and for varying types of assistance. The informal support system of kin, friends, and neighbors constitutes the most involved circle in daily life. Older adults turn to the members of an informal system, from inner to exterior, which are available to provide assistance to them.

From the perspective of Chinese customs, traditional familism advocates an individual's devotion to family members and the benefits brought to the family, as well as the offspring. Additionally, the universally recognized collectivism among the Chinese ethnic group impels Chinese to care for and rely on family members. However, the norms have been undergoing delicate changes in this rapidly developing society, for example, Hong Kong is now one of the leading regions in terms of a high degree of modernization. As a result, people tend to unearth more available resources. Thus, for the older adults living in the community, surrounding peers could constitute a type of support network through which social support is exchanged reciprocally. Nevertheless, it is found that excessive interactions with friends could diminish well-being because of collectivistic norms that discouraging the expression of one's personal concerns, which may signify problems in the group or may disrupt harmony between friends. Therefore, emotional communication with peers is not accorded much value, but it makes no difference in terms of companionship because there is less emotion and privacy permeation in solely accompanying others. This interpretation sheds light upon the pursuit of emotional closeness in later life and further leads to the differentiation of emotional and companionship support in the process of emotional selectivity. However, the limited questions in both the survey questionnaires and interview guidelines imposed restriction on the examination of the inner demands and reciprocal behaviors of the two types of support. Therefore, the present study could not show a full picture of this issue, but could only exhibit an abbreviated view.

9.4 The Effect of Social Support Networks on Positive Aging

The present study examined the structure and functions of the convoy model, with a research focus on the family network and peer network, as well as their ability to predict the effects of positive aging. The findings of hierarchical regression analysis tended to support the influence of social support networks on positive aging.

9.4.1 *The Contribution of Family and Peer Support Networks to Positive Aging*

The results of the regression analysis showed that family network size predicted a positive effect on the overall positive aging and its three components—CE, PE, and life satisfaction after age, gender, and perceived socioeconomic status have been controlled for. Additionally, the findings, in terms of functional family support, reflected a positive effect on the overall positive aging, CE, and PE. Furthermore, peer network size was found to have a positive effect on the overall positive aging and all its five components. Finally, the functional peer support positively predicted the overall positive aging and other four components, except for CE.

Numerous studies have found a positive relationship between social support networks and psychological well-being (Poulin, Deng, Ingersoll, Witt, & Swain, 2012), or aging well (Adams & Blieszner, 1995), among older adults. Two primary sources of interpersonal relationships to acquire emotional closeness (Carstensen et al., 1999), family members and peer groups, have been suggested to predict well-being in later life. The findings, in terms of the effect of both family network (size and functions) and peer network (size and functions) on the overall positive aging, are similar to the previous results. The present findings lend support to the empirical findings that family and peer networks are vital resources for supporting older persons to age well (Adams & Blieszner, 1995; Cheng et al., 2009). Additionally, the positive contribution of family and peer networks to one component of positive aging—life satisfaction—reveals their significant role in enhancing subjective well-being, which is in line with the implications of empirical studies (Phillips, Siu, Yeh, & Cheng, 2008).

A unique finding discovered in this study supports the fact that peer network size makes more contributions to the components of positive aging than family network size. Statistics showed that family network size did not significantly predict two components of positive aging, illness avoidance or functioning, while peer network size did. Since the present study could not provide the causal interpretation of the existing relationships between two parties, I tried to elucidate the findings. It has been recorded that family members and peers can provide resources for the older people in two ways—by helping them instrumentally and by providing emotional support (Adams & Blieszner, 1995). In Chinese society, primary family members usually constitute the major source of elderly caregiving, so that older adults tend to rely on specific family members if they are in poor health. However, with the exception of spouses, most family members are not chosen. Therefore, the total size of family networks remains relatively constant even though the older person might be experiencing deterioration in health. This is probably the reason why the relationships between family network size and health and functioning are ambiguous. The positive effect of peer network size on health and functioning could not be precisely elaborated in this study by virtue of existing evidence and knowledge. Nevertheless, the positive association between the variables could be interpreted from the nature and feasibility of the development of peer relationships among older adults. Researchers have found that older adults in better health condition tend to have more friendship (Hartup

& Stevens, 1997). Consequently, it is reasonable to infer that the participants in the present sample who have less illness and better functioning ability are prone to develop more friendships within peer groups. This is a potential reason for the association between peer network size and health and functioning. Based on the above inference, further studies aiming to explore the causal relationship between family and peer network size and its effect on the components of positive aging are suggested. Future studies are also suggested to examine the contact frequency and reciprocal behaviors within family and peer networks, and to identify the way through which family and peer networks contribute to positive aging in later life.

Similarly, the effect of functional family support and functional peer support on overall positive aging and its components could be uncovered. It should be noticed that functional family support was not associated with life satisfaction, while functional peer support was positively associated with life satisfaction. This evidence is in line with a body of empirical findings indicating that friends are more important to the psychological well-being of older adults than family members (Antonucci & Akiyama, 1995; Birditt, Jackey, & Antonucci, 2009) because of fewer negative interactions exist within peer groups than among family members. As a matter of course, this result is also consistent with the prevalence of peer support derived from the findings of social network composition analysis and the in-depth interview, which was discussed earlier in this chapter.

9.4.2 The Interpretation of Emotional Closeness and Its Effect on Positive Aging

In addition to family and peer network, emotional closeness, as a core concept of socioemotional selectivity theory, has been claimed to affect well-being positively in previous studies (Antonucci & Akiyama, 1987a, 1987b; Antonucci, Fiori, Birditt, & Jackey, 2010). To explore emotional closeness among Chinese older adults in the present sample, hierarchical regression analysis was conducted to test its predictive effect on positive aging. The statistics showed that respondents with stronger family and peer emotional closeness reported better achievement of positive aging after age, gender, education, and perceived SES were controlled for. Meanwhile, integrated emotional closeness (operationalized as the sum of family emotional closeness and peer emotional closeness) was also found to have a positive effect on PA. This finding is consistent with the postulation derived from socioemotional selectivity theory that emotional closeness is beneficial to well-being in later life (Carstensen, 1992; Fung et al., 2008). Moreover, the finding confirms the importance of emotional closeness in contributing to aging well among Chinese older adults, which further expands the applications of socioemotional selectivity theory.

Given the evidence that emotional closeness in this study was found to positively predict positive aging, two distinctive implications could be unveiled from the findings. First, it could be inferred that the contribution to positive aging would depend on more resources from the total network, transmitted through help and support, than from close social networks. That is, resources existing in the total social network would be more prevalent than perceived close or significant social partners. Regression analysis showed that the number of close social partners had a positive effect on the overall positive aging after controlling for age, gender, education, and perceived SES. However, in the second step, the effect of close social network size on positive aging disappeared after the total network size was controlled for. The spurious results trigger further speculation regarding the role of close social network size in comparison to total social network size and their predictive effects of positive aging. The two-step regression analysis, in terms of close social network size and total network size, showed that the total network size was more predictive of positive aging, regardless of how many close social partners were in the social network. This finding is also in line with the result of a previous study conducted by Chong, Cheung, Woo, and Kwan (2012), who argued that the larger the network size, the higher the well-being among Hong Kong Chinese older adults.

However, the weak prediction of close social network size to positive aging calls for more elaboration. The results of this study show that the number of close social partners might not provide adequate and required resources and help. This implication may result from the nature of close relationships. The close social network usually consists of intimate kin or friends, who cannot be chosen freely on most occasions. Also, the number of close social partners generally remains constant at a specific stage in one's lifespan. Therefore, unpleasant interactions with close partners and the negative relational consequences of help-seeking brought by traditional norms of reciprocity may hamper the achievement of positive aging (Chong et al., 2012). Hence, having more close social partners is not necessarily associated with better positive aging outcomes.

Second, it should be postulated that close network size is different from emotional closeness in predicting positive aging. This finding may trigger a revisitation of the implications and the framework of emotional closeness. In past studies, emotional closeness was primarily researched from the perspective of its depth, namely, the quality of close relationships or close interaction and intimacy with close social partners. Moreover, another similar indicator in relation to emotional closeness may be the breadth of the close social network, which concerns the number of close social partners that the focal person perceives. To the best of our knowledge, few studies were found to examine the relationship between the breadth of close social networks and the depth of emotional closeness. Although the authoritative definition of emotional closeness seems unidentified, it is generally accepted that emotional closeness between two people leads them want to help each other, especially when one is in need (Korchmaros & Kenny, 2001). Emotional closeness entails the sense of concern, trust, and caring for another individual and the enjoyment of the relationship with that individual (Cunningham, 1986; Lee, Mancini, & Maxwell, 1990).

With reference to previous statements, defined from the attribute of depth, emotional closeness in the present study was operationalized as the degree of indicators showing a willingness to interact or discuss intimate feelings with others. As for the breadth of a close social network, it is defined as the number of close social partners, operationalized as the number of social partners falling in the inner circle of the Social Convoy Questionnaire. The statistics showed that the breadth of close social networks had fewer impacts on positive aging, and, thus, did not as contribute to the depth of emotional closeness, namely, close interaction and intimacy with close social partners. Additionally, correlation analysis revealed that the number of close social partners was not significantly associated with emotional closeness. Therefore, these findings tend to highlight the effect of the depth of emotional closeness on positive aging. This implication probably supports the view that the depth of emotional closeness encompasses more essential elements (e.g., love, caring, trust) planted in the process of interactions with close social partners.

To summarize, the breadth of close social networks should not be treated equivalently to the depth of emotional closeness in determining their predictive effects on positive aging or other well-being issues, though the former has been considered a prerequisite for the latter.

It is a limitation that the present study did not further examine the quality of interactions and contact frequencies with close social partners. Therefore, the link between the breadth of close social partners and the depth of emotional closeness remains unknown. In view of the above inference, future studies designed to examine with whom emotional closeness is more inclined to be cultivated and the way through which the depth of emotional closeness could facilitate positive aging or well-being are recommended.

9.5 The Significant Functions of Social Support

Four types of functional social support—emotional support, instrumental support, companionship support and informational support—were researched in the present study. Descriptive statistics showed that the mean of received emotional support ranked highest, followed by companionship support. In terms of the changes of receiving different functional support within the last ten years, the findings revealed that respondents received more functional social support than ten years ago, particularly, informational support.

The findings, in terms of the prevalence of emotional support and companionship support, further addressed their important role in the later life of community-dwelling older adults. It is also consistent with the viewpoint of socioemotional selectivity theory, that is, older adults are inclined to receive more emotional help and companionship because of their motivation to pursue emotional closeness (Carstensen et al., 1999; Lang & Carstensen, 2002). As to the changes of receiving social support, the present findings are in line with empirical evidence claiming that older adults expect to receive more social support as they are getting older (Barrera & Ainlay, 1983; House, Umberson, & Landis, 1988). A number of studies show that the elderly tend to need more social support, although they actually receive less than before, because of the decreasing social resources and decreasing involvement in social interactions (Depner & Ingersoll-Dayton, 1988). The implication derived from this study suggests that participants living in the community in this sample possess more resources that could provide social support in need. Additionally, with reference to the previous discussion on the contribution of total network size to positive aging, the present findings also support the viewpoint that available resources in regard to giving help and support, which are embedded in social networks, play a crucial role in community life.

The increasing receipt of informational support among the community-dwelling older adults reflects the increasing demand of getting equipped with information in this knowledge-based society. The findings showed that among the five types of social networks, professionals and primary family members made great contributions to provide informational support. Furthermore, means tests revealed that professionals (e.g., social workers, nurses, doctors, physicians) brought in the latest information and knowledge to the elderly more than primary family members did. In general, informational support, by contrast, is not perceived as helpful by all sources. Helgeson (2003) claims that people are not always fond of informational support provided by family and friends. In some occasions this would be viewed as unwanted advice, although people may want informational support from experts, e.g., from nurses and physicians. The findings in the present study confirmed the important role of professionals in giving informational support, which is consistent with the argument of Helgeson (2003). However, the informational support provided by family members is still valued by the older Chinese people in this sample.

Krause (1986) argues that the utility of informational support is helpful by providing people with knowledge that they can use to resolve a stressful situation. Hong Kong is on the way to embracing the rapid development of information and technology. The changing situation also brings in opportunities and platforms for older adults to keep updated with recent trends. Additionally, older adults may need more caring and help in society. They should have access to relevant information on social insurance, social security, medical care, lifelong learning, and so on. Therefore, the receipt of informational support is gradually increasing with the development of society. During the interview process, respondents mentioned the role of professionals (e.g., social workers, doctors, nurses) in providing information that they required. Thus, I can infer that the increasing receipt of informational support may also attribute to information transmission by professionals, from whom the community-dwelling older people get informed.

9.6 Coping in Later Life and Its Mediating Effects on Social Support Networks and Positive Aging

In this section, a discussion regarding the coping strategies applied by the community-dwelling older adults and its mediating effect on social support network and positive aging will be presented. In the process of data analysis related to the issues in this section, one-way ANOVA was conducted to test for differences in three types of coping strategies among the three age groups (e.g., the young-old, mid-old and old-old). The findings revealed that significant differences existed among the three age groups. Additionally, path analysis was performed to test the interrelationships between the main variables. Among all the trials of modeling, peer emotional closeness was verified to partially mediate the relationship between social support networks and positive aging. This section will first discuss the implications in terms of the three types of coping strategies among three age groups in this sample. Second, inferences based on the findings in relation to the mediating role of coping will be elaborated.

9.6.1 Age and Coping

Compared with the young-old, the old-old reported lower applications of problem coping and emotion coping strategies. Regarding avoidance coping, no significant difference was detected between the three age groups, although avoidance coping ranked the highest among the three types of coping strategies. This fact is in accordance with empirical findings. However, the decreasing application of problem and emotion coping as people aged is not entirely consistent with the empirical findings. The developmental hypothesis suggests that when people get older, certain inherent, stress-related changes occur, resulting in use of either more mature or more primitive coping options (Folkman, Lazarus, Pimley, & Novacek, 1987). Additionally, another similar argument of Folkman and associates suggests that older people may use more passive, emotion-focused coping than the younger group. Another study also showed that age differences existed in problem-solving styles, with older adults (65–75 years) using less problem-focused coping and more passive coping than young or middle-aged adults (Blanchard-Fields, Jahnke, & Camp, 1995). However, the above evidence is mostly from western countries, thus it might not apply to Chinese societies. Older Chinese people were indoctrinated with traditional social norms and values under a collectivistic culture. Thus, when older Chinese adults are confronted with stresses or troubles, they sometimes tend to avoid seeking help or advice from others, which is reflected in their less frequent use of problem and emotion coping. This tendency could be explained by the fact that in traditional Chinese culture, a person should be self-sufficient and seeking help may be deemed as a failure (Chong et al., 2012; Litwin, 2010).

Another implication might be derived from the present findings in terms of the preference of coping strategies. It has long been accepted that coping represents the cognitive and behavioral efforts of individuals to manage stressful encounters (Folkman et al., 1987). When stressful events occur, people may stimulate actions or emotions to cope with the stresses. The interpretation of the actual situation of coping probably goes to the nature of the sample. It should be noticed that the participants in this sample are all community-dwelling older adults as members in a community center. All participants consented to participate in the survey. Thus, it could be inferred that the respondents generally have an agreeable or delightful mood in normal times, or at least during the time of being interviewed. Hence, participants in this study might report fewer stresses or worries than other group of older adults. This resulted in more "don't know" answers in regard to coping strategies, that is, they didn't think they needed to use coping strategies because they could hardly recall specific stressful events. This phenomenon is more prevalent among the old-old because they expressed severe resistance to talking about stresses and worries in life. This could be a major reason why the old-old tend to use fewer problem coping and emotional coping strategies. Nevertheless, the age difference in the three age groups, in terms of coping strategies in this sample, remains unresolved. Further exploration (e.g., qualitative studies) is suggested to uncover whether old people at different ages have the habit of employing certain specific types of coping strategies.

9.6.2 The Mediating Effect of Coping on Social Support Network and Positive Aging

The statistics only support one integrated path model of positive aging. In this model, both problem coping and emotion coping served as a mediator between peer emotional closeness and positive aging. Additionally, path analysis showed that problem coping could separately mediate the effect of family networks on positive aging. Also, emotion coping could separately mediate the effect of peer networks on positive aging. In examining the effect of emotional closeness (e.g., peer emotional closeness and family emotional closeness) on positive aging, problem coping was found to serve as a mediator in the path model. Similarly, emotion coping also mediated the effect of emotional closeness on positive aging.

The joint mediating effect of problem coping and emotion coping on peer emotional closeness and positive aging reveals that peer emotional closeness could trigger problem coping and emotion coping at a proper time. At the same time, problem coping and emotion coping could contribute to the achievement of positive aging. Additionally, the significant indirect effect of peer emotional closeness, via problem coping, on positive aging reflects that the use of problem coping strategies plays an important role in contributing to positive aging. Thus, it is worth noting that close peer groups could provide resources or assistance to older adults to help them tackle problems in life. Furthermore, the marginally significant indirect effect of peer emotional closeness, via emotion coping, on positive aging shows that close peer groups could also trigger emotion coping, which is positively predictive of positive aging, though this indirect effect is reluctantly accepted.

Although statistics only verified one integrated path model combining problem coping and emotion coping, an alternative explanation might be the collinearity between problem coping and emotion coping. In the integrated path model, the evidence is particularly clear in regard to problem coping as a mediator, revealing the prevalence of problem coping among the older adults in the current research. At the same time, the importance of close peer groups is reconfirmed through the path model, which is consistent with the former findings of the present study.

9.7 Summary

To summarize, this chapter has elucidated the implications of findings on social networks, coping, and positive aging in four main aspects. First of all, the study validated an expanded Positive Aging Index adjusting to the older adults living in community. The 19-item scale was verified as a tool to examine positive aging under the Chinese cultural context. Second, social support from peer groups was found to positively predict positive aging. Peer groups in later life play a significant role among community-dwelling, older Hong Kong Chinese. Third, companionship, as an attribute of functional support, is mostly valued by community-dwelling, older Chinese. Leisure activities and daily communications within the context of elderly centers and whole communities have become resources for providing companionship to older adults, which is also associated with better achievement of positive aging. Last but not least, problem and avoidance coping were the most commonly used strategies in this sample, especially for the old-old. Additionally, problem coping and emotion coping jointly mediated the relationship between peer emotional closeness and positive aging. This finding confirms the mediator role of coping in the predictive effect of emotional closeness on overall positive aging.

References

Adams, R. G., & Blieszner, R. (1995). Aging well with friends and family. *American Behavioral Scientist, 39*(39), 209–224.

Antonucci, T. C. (2001). Social relations: An examination of social networks, social support, and sense of control. In J. E. Birren (Ed.), *Handbook of the psychology of aging* (5th ed., pp. 427–453). San Diego, CA: University of California; Center on Aging.

Antonucci, T. C., & Akiyama, H. (1987a). An examination of sex differences in social support in mid and late life. *Sex Roles, 17*(11), 737–749.

Antonucci, T. C., & Akiyama, H. (1987b). Social networks in adult life and a preliminary examination of the convoy model. *Journal of Gerontology, 42*(5), 519–527.

Antonucci, T. C., & Akiyama, H. (1995). Convoys of social relations: Family and friendships within a life span context. In R. Blieszner & V. H. Bedford (Eds.), *Handbook of ageing and the family* (pp. 355–371). Westport, CT: Praeger Publisers.

Antonucci, T. C., Fiori, K. L., Birditt, K., & Jackey, L. M. (2010). Convoys of social relations: Integrating life-span and life-course perspectives. In R. M. Lerner, M. E. Lamb, & A. M. Freund (Eds.), *The handbook of life-span development* (Vol. 2, pp. 434–473). Hoboken, NJ: Wiley.

Barrera, M., & Ainlay, S. L. (1983). The structure of social support: A conceptual and empirical analysis. *Journal of Community Psychology, 11*(2), 133–143.

Birditt, K. S., Jackey, L. M. H., & Antonucci, T. C. (2009). Longitudinal patterns of negative relationship quality across adulthood. *The Journals of Gerontology, Series B: Psychological Sciences and Social Sciences, 64*(1), 55–64.

Blanchard-Fields, F., Jahnke, H. C., & Camp, C. (1995). Age differences in problem-solving style: The role of emotional salience. *Psychology and Aging, 10*(2), 173–180.

Bosworth, H. B., & Schaie, K. W. (1997). The relationship of social environment, social networks, and health outcomes in the Seattle Longitudinal Study: Two analytical approaches. *The Journals of Gerontology, Series B: Psychological Sciences and Social Sciences, 52*(5), 197–205.

Bowling, A. (2007). Aspirations for older age in the 21st century: What is successful aging? *International Journal of Aging and Human Development, 64*(3), 263–297.

Cantor, M. H. (1979). Neighbors and friends: An overlooked resource in the informal support system. *Research on Aging, 1*(4), 434–463.

Carstensen, L. L. (1992). Social and emotional patterns in adulthood: Support for socioemotional selectivity theory. *Psychology and Aging, 7*(3), 331–338.

Carstensen, L. L., Isaacowitz, D. M., & Charles, S. T. (1999). Taking time seriously: A theory of socioemotional selectivity. *American Psychologist, 54*(3), 165–181.

Cheng, S. T., Lee, C. K. L., Chan, A., Leung, E. M. F., & Lee, J. J. (2009). Social network types and subjective well-being in Chinese older adults. *The Journals of Gerontology, Series B: Psychological Sciences and Social Sciences, 64*(6), 713–722.

Chong, A. M. L., Cheung, C. K., Woo, J., & Kwan, A. Y. H. (2012). Availability, use, and cultivation of support networks as predictors of the well-being of middle-aged and older Chinese: A panel study. *The Scientific World Journal, 2012*.

Chong, A. M. L., Ng, S. H., Woo, J., & Kwan, A. Y. H. (2006). Positive ageing: The views of middle-aged and older adults in Hong Kong. *Ageing & Society, 26*(2), 243–265.

Crohan, S. E., & Antonucci, T. C. (1989). Friends as a source of social support in old age. In R. G. Adams & R. Blieszner (Eds.), *Older adults friendship: Structure and process* (pp. 129–146.) Newbury Park, CA: Sage.

Cunningham, M. R. (1986). Levites and brother's keepers: A sociobiological perspective on prosocial behavior. *Humbolt Journal of Social Relations, 13*(1/2), 35–67.

Deng, J., Hu, J., Wu, W., Dong, B., & Wu, H. (2010). Subjective well-being, social support, and age-related functioning among the very old in China. *International Journal of Geriatric Psychiatry, 25*(7), 697–703.

Depner, C. E., & Ingersoll-Dayton, B. (1988). Supportive relationships in later life. *Psychology and Aging, 3*(4), 348–357.

Depp, C. A., & Jeste, D. V. (2006). Definitions and predictors of successful aging: A comprehensive review of larger quantitative studies. *American Journal of Geriatric Psychiatry, 14*(1), 6–20.

Fisher, B. J. (1995). Successful aging, life satisfaction, and generativity in later life. *The International Journal of Aging and Human Development, 41*(3), 239–250.

Folkman, S., Lazarus, R. S., Pimley, S., & Novacek, J. (1987). Age differences in stress and coping processes. *Psychology and Aging, 2*(2), 171–184.

Fung, H. H., Stoeber, F. S., Yeung, D. Y. L., & Lang, F. R. (2008). Cultural specificity of socioemotional selectivity: Age differences in social network composition among Germans and Hong Kong Chinese. *The Journals of Gerontology, Series B: Psychological Sciences and Social Sciences, 63*(3), 156–164.

Hartup, W. W., & Stevens, N. (1997). Friendships and adaptation in the life course. *Psychological Bulletin, 121*(3), 355–370.

Helgeson, V. S. (2003). Social support and quality of life. *Quality of Life Research, 12*(Suppl. 1), 25–31.

House, J. S., Umberson, D., & Landis, K. R. (1988). Structures and processes of social support. *Annual Review of Sociology, 14*(1), 293–318.

Kahn, R. L., & Antonucci, T. C. (1980). Convoys over the life course: Attachment, roles and social support. In P. B. Baltes & O. Brim (Eds.), *Life-span development and behavior* (Vol. 4, pp. 253–286). New York: Academic Press.

Korchmaros, J. D., & Kenny, D. A. (2001). Emotional closeness as a mediator of the effect of genetic relatedness on altruism. *Psychological Science, 12*(3), 262–265.

Krause, N. (1986). Social support, stress, and well-being among older adults. *Journal of Gerontology, 41*(4), 512–519.

Lang, F. R., & Carstensen, L. L. (1994). Close emotional relationships in late life: Further support for proactive aging in the social domain. *Psychology and Aging, 9*(2), 315–324.

Lang, F. R., & Carstensen, L. L. (2002). Time counts: Future time perspective, goals, and social relationships. *Psychology and Aging, 17*(1), 125–139.

Le, G., Shen, J., & Du, L. (2004). *The development of social psychology in China*. Tianjin, China: Tianjin People. (in Chinese).

Lee, T. R., Mancini, J. A., & Maxwell, J. W. (1990). Sibling relationships in adulthood: Contact patterns and motivations. *Journal of Marriage and the Family, 52*(2), 431–440.

Litwin, H. (2010). Social networks and well-being: A comparison of older people in Mediterranean and non-Mediterranean countries. *The Journals of Gerontology, Series B: Psychological Sciences and Social Sciences, 65*(5), 599–608.

Merz, E. M., & Huxhold, O. (2010). Wellbeing depends on social relationship characteristics: Comparing different types and providers of support to older adults. *Ageing and Society, 30*(5), 843–857.

Ng, S. H., Cheung, C. K. J., Chong, A. M. L., Woo, J., Kwan, A. Y. H., & Lai, S. F. Y. (2011). Aging well socially through engagement with life: Adapting Rowe and Kahn's model of successful aging to Chinese cultural context. *The International Journal of Aging and Human Development, 73*(4), 313–330.

Phillips, D. R., Siu, O. L., Yeh, A. G. O., & Cheng, K. H. C. (2008). Informal social support and older persons' psychological well-being in Hong Kong. *Journal of Cross-Cultural Gerontology, 23,* 39–55.

Pinquart, M., & Sorensen, S. (2000). Influences of socioeconomic status, social network, and competence on subjective well-being in later life: A meta-analysis. *Psychology and Aging, 15,* 187–224.

Poulin, J., Deng, R., Ingersoll, T. S., Witt, H., & Swain, M. (2012). Perceived family and friend support and the psychological well-being of American and Chinese elderly persons. *Journal of Cross-Cultural Gerontology, 27*(4), 305–317.

Pruchno, R. A., Wilson-Genderson, M., & Cartwright, F. (2010). A two-factor model of successful aging. *The Journals of Gerontology Series B, 65*(6), 671–679.

Rowe, J. W., & Kahn, R. L. (1987). Human aging: Usual and successful. *Science, 237*(4811), 143.

Rowe, J. W., & Kahn, R. L. (1997). Successful aging. *The Gerontologist, 37*(4), 433–440.

Ryff, C. D. (1982). Successful aging: A developmental approach. *The Gerontologist, 22*(2), 209–214.

Chapter 10
Implications of the Study

This chapter is divided into three main parts. The first proposes recommendations to support the knowledge, policies, services and research implications discussed in the findings of this study. The second part discussed the limitations of the research design. Finally, an exanimation of possible directions for future research follows.

10.1 Recommendations

As shown in Chap. 9, the presence of a social support network and knowledge of coping strategies were found to have a strong positive effect in older Hong Kong Chinese living in the community. This section presents recommendations derived from the study in regard to four areas: (1) promoting positive aging in Chinese societies, (2) strengthening elders' social support network, (3) cultivating coping skills, and (4) improving the service quality of domestic helpers.

10.1.1 Promoting Positive Aging in Chinese Societies

Though the present study concerned older Hong Kong Chinese, the conceptual framework and key determinants of positive aging are likely to enhance understanding of the elderly in other Chinese communities such as Taiwan and cities in Mainland China. Positive aging has received attention in Hong Kong in recent years, both in the academic and social welfare areas. Additionally, different age cohorts across various societies have distinct perspectives on positive aging. As such, the notion of positive aging should be promoted in other Chinese societies. This would help to identify additional characteristics of positive aging from the perspectives of middle-aged adults in various regions. It is also recommended that other normatively desirable states of positive aging, such as positive spirituality (e.g., Crowther, Parker, Achen-

S. Liu, *Social Support Networks, Coping and Positive Aging Among the Community-Dwelling Elderly in Hong Kong*,
https://doi.org/10.1007/978-981-13-3783-3_10

baum, Larimore, & Koenig, 2002), be examined in a wider scope of the Chinese population. To yield even more insight, closely related concepts such as healthy aging and active aging should be explored in conjunction with positive aging.

Promoting Productive Engagement

Implications from empirical studies on positive aging have led to the formation of policies regarding productive engagement a (Chong, Ng, Woo, & Kwan, 2006; Ng, Woo, Kwan, & Chong, 2006). For example, *Senior Volunteerism* is highly promoted by the DECC as a manifestation of productive engagement. It aims to encourage elders to become volunteers, to be continuously involved in social affairs, and to serve other elders in need so as to promote a sense of worthiness (SWD, 2014). In addition, a program titled *Opportunities for the Elderly Project (OEP)* was launched in 1998 by the SWD to promote a sense of worthiness among the elderly in Hong Kong (SWD, 2014). This program has become a regular service of the SWD and has received support from the Lotteries Fund for more than ten years. The OEP encourages the elderly to actively take part in community affairs and to make good use of their expertise and ample experience in life by offering their time to various social service organizations, district organizations, and educational institutes. Also, the OEP encourages other stakeholders (e.g., churches, schools, or NGOs) to promote a sense of worthiness in older individuals.

The current measures that exist to facilitate productive aging are suggested to be strengthened to attract and retain older people as well as mid-life adults; this would encourage volunteerism from a life-span perspective and further promote productive engagement. Community education can be used to promote a positive image of older adults and advocate that older adults, including those with disabilities or illnesses, can contribute to society as volunteers. In addition, in order to appeal to older people interested in devoting time to volunteering, a favorable attitude toward volunteerism among older people and their families should be nurtured through community or organizational education.

Well-designed training programs may be offered in order to help older volunteers master the skills and knowledge needed to render voluntary service; such programs would also enhance their perceived competence. District councils are advised to devote more effort to promoting volunteerism among the elderly through, for example, formulating district training plans and incentive measures and attracting more funding to invest in elder volunteerism. Organizations interested in elder volunteering can cooperate with districts by virtue of the wide distribution of district board members. Apart from the DECC, other nonprofit organizations (e.g., Neighborhood Elderly Center [NEC] and Social Center for the Elderly [SCE]) could provide more opportunities for the elderly to participate in volunteer activities.

In terms of facilitating other types of productive engagement, it is recommended that SCEs organize more social and recreational activities for elders in the community to help them make constructive use of their leisure time, develop their potential through learning, and build social networks.

Promoting Caring Engagement

In 2009, in order to encourage caring engagement among the elderly, the Hong Kong Housing Society (HKHS) introduced the *Harmonious Families Priority Scheme* to help care for elderly parents or dependents and promote harmony in the family. With this scheme, eligible families may opt to live in one or two nearby flats, according to their choice of district and the number of flats available for their family situation. This makes it more convenient for adult children to look after their parents, especially those who are in need of caregiving (Hong Kong Housing Authority 2014). This scheme supports the meaning of caring engagement of positive aging, encourages more interactions and caring within generations, and helps the elderly lead a happier life through living with or nearby their offspring.

It is suggested that communities should organize more intergenerational activities and nurture a harmonious atmosphere between older adults and the younger generation. In addition, efforts should be made to stimulate the enthusiasm of adult children to have healthy, fulfilling relationships with their older parents (e.g., creating more opportunities for adult children to communicate with their older parents and listen to their thoughts in order to understand their needs and wishes). Consequently, older adults who have satisfactory interactions with younger generations will be more devoted to caring engagement because of the effect of reciprocity.

The importance of contributing to the family should be brought to the attention of the elderly through mass media or community education, emphasizing the various forms of caring engagement. In addition to household maintenance activities and taking care of grandchildren, the elderly are encouraged to give emotional support to younger generations in need. Older adults have experienced lots of events in life, making them uniquely qualified to offer advice or help to younger generations; this is also deemed to facilitate caring engagement.

Apart from the recommendations regarding caring engagement and productive engagement, measures should be taken to promote physical and functional health among the elderly. The findings of the present study reveal that good health status constitutes an important part of positive aging, which is a prerequisite for caring and productive engagement. Hence, improvement in health care services and the medical system are recommended. *Health Care Voucher*, a health care scheme recently released by the Hong Kong Government, encourages older people to visit private sector services and provides a sum of money to subsidize their use. By virtue of this program, adults over the age of 70 would gain resources to consult private medical services and be able to utilize preventive health care services more often. The scheme collaborates with Chinese medicine practitioners, medical practitioners, dentists, chiropractors, optometrists and physiotherapists to provide medical and health care to the elderly. According to the Interim Review of Elderly Health Care Voucher Pilot Scheme (2011), only 35% of the interviewed elderly actually used the vouchers, indicating a large pool of potential users. Therefore, it is recommended that policy makers work to attract more service providers to join the scheme and to adjust the amount of the health care voucher for the benefit of older adults.

10.1.2 Investing in and Consolidating the Social Support Network of the Elderly

In addition to the contributions of the family network to positive aging, the current study has revealed the importance of the role of peers in providing companionship support to the community-dwelling elderly. Moreover, the findings have shown that the total social network size and emotional closeness (i.e., family and peer emotional closeness) were positively associated with positive aging. Hence, it is beneficial to encourage the elderly to develop and invest in a solid social support network as a convoy to provide social support.

It is recommended that elderly individuals living in the community develop more social relations with neighbors, old friends, center members, and others with whom they can mutually share emotions or companionship. In light of the notion of *aging in place* (AiP), older adults are expected to stay in the community as long as they wish and lead a healthy and productive later life. The community surroundings provide not only relevant infrastructure resources (e.g., fitness facilities, disabled access and frail elderly access) but also potential social network resources.

Many older adults live separately from their adult children who are busily engaged in their own career and family. As a result, the community-dwelling elderly are encouraged to leave home and identify appropriate partners or friends to form a peer support network. As a result, older adults are encouraged to participate in activities or attend social events with peer group members, enabling them to obtain happiness and fulfillment. In several DECCs of Hong Kong, a number of older volunteers are trained to provide support to other peer members with lower socioeconomic status or special needs (e.g., solitude because of living alone). They regularly visit the widowed or those who live alone, deliver food to them and present amusing performances. This productive and reciprocal program is worth promoting, as it allows older adults to benefit from mutual help and contributes to a consolidated peer group network.

As shown in previous chapters, this study found that emotional closeness is a significant indicator to predict positive aging. It is not uncommon that having close ties with family members often leads to negative interactions and unavoidable responsibilities. Nonetheless these close relationships are still highly valued. Hence, it is imperative for older adults to cultivate their network of close social partners with whom they can discuss emotions and needs. It is recommended that older adults sympathetically and genuinely manage the close relationships they have with family member in order to maintain a satisfying relationship. They are advised to spend more time communicating with close family members and to be tolerant and supportive toward them. For the close ties with peer group members, older adults are recommended to actively make friends with peers, maintaining stable and reciprocal relationships. For the elderly living in the community, positive interrelationship with one's peer group helps to expel loneliness and reduce reliance on other family members. Participation in volunteering work and attending religious activities are good ways to get to know more friends and selectively form close relationships. For example, elderly individuals could have regular contact with volunteering partners or church and center members, ensuring that they are available to develop friendships. They can also seek interactions with others sharing similar hobbies and interests (e.g., sports and leisure activities).

10.1.3 Improving Coping Capacity in Later Life

The present study found that the use of problem and avoidance coping were more prevalent among community-dwelling elderly in Hong Kong. Problem coping generally leads to actions that can save energy and efforts, thus eventually resulting in alternative rewards. In contrast, avoidance does not aim to solve problems. Findings also revealed that family network and peer network were positively predictive of problem coping and emotion coping. Therefore, it is recommended that the elderly cultivate coping strategies in order to facilitate positive aging in later life. First, maintaining a stable and positive relationship with family members and peer groups is recommended. Older people have likely experienced similar life events or transitions, and so they might be more willing to share the feelings with peer groups. The act of mutual sharing has been found to play a significant role in providing emotional and companionship support to older adults, serving as a stimulus to well-being. Second, avoidance is a common way older people cope with stress. It is recommended that the elderly adopt appropriate avoidance coping strategies even though avoidance coping cannot actually help with problem-solving. In old age, people may not have the strong motivation or resources to solve problems or to overcome difficulties because they tend to pay less attention to affairs concerning future goals (Carstensen, Isaacowitz, & Charles, 1999). Therefore, avoidance coping is adopted in later life, as it helps relieve stressors. However, avoidance coping is sometimes not considered an effective coping strategy because it is related to less favorable outcomes and negative effects or adjustments (Bowman & Stern, 1995). Hence, it is suggested that older adults adopt appropriate coping strategies (e.g., letting go, comforting oneself, accepting the reality) that account for situational factors (e.g., the necessity to solve problems). Though there are not consistent findings showing patterns of coping in later life, older adults should master some efficient coping strategies, which could be tailored in accordance with their characteristics and backgrounds.

10.1.4 Improving the Service Quality of Domestic Helpers

According to statistics, foreign domestic helpers (FDHs) have accounted for a large share of the caregivers serving older adults in the community of Hong Kong. In view of the findings showing that only 40% of those who employed domestic helpers reported satisfactory or very satisfactory services, proper measures should be carried out to improve the quality of service. The first recommendation is to strengthen the relationship between older adults and FDHs. Reports from the open questions regarding the disadvantages of employing FDHs showed that some older adults did not get along well with their FDHs, which in turn led to unsatisfactory service ratings. Hence, both parties are advised to cultivate better communication with each other and be considerate of the other's customs and habits. In addition, care recipients are advised to list the requirements of caregiving and standards of services for their FDHs; this is beneficial to forming a better employer–employee relationship. In addition, primary family caregivers of older adults should coordinate with FDHs and pay attention to the needs and feelings of both the care recipients and their FDHs.

Additionally, employment agencies should provide training programs for FDHs taking care of older adults in order to fulfill the caregiving needs of frail elderly living in the community. A number of care recipients are physically weak or suffering from dementia, which calls for professional and reliable care. In such cases, FDHs who are going to take care of frail older adults are advised to obtain relevant professional knowledge and skills in caregiving.

10.2 Limitations

There are several limitations in the research design of the current study. The first lies in the restriction of the sampling method in generalizing the findings to the total older population in Hong Kong. Despite concerted efforts to recruit a diverse sample from 18 districts of Hong Kong, only a small proportion of community-dwelling older adults participated in the study. Additionally, the profile of the participants revealed that older adults who had a network of center members might be over-represented in the sample. This means that community-dwelling older adults who are not members of community elderly centers may under-represented. Therefore, to some extent, the prevalence of peer support found in the present study could be attributed to the nature of the sample and cannot be generalized to other elders in Hong Kong.

Second, the research design was cross-sectional; hence we were unable to assess causality and to infer cause-effect relationships. Results of path analysis showed that coping (e.g., problem coping and emotion coping) served as a mediator in the relationship between peer emotional closeness and overall positive aging. Nevertheless, it should be noted that the reverse effect is equally possible; for example, positive aging may have generated an effect on coping, emotional closeness, and so on. Moreover, because our findings elucidate the interrelationships among variables at a single point of time, they do not enable understanding of the process or trajectories of positive aging from a lifespan perspective.

The third limitation relates to the measurement of the dependent variable. Although the measurement of positive aging (i.e., expanded PAI) has been further validated among Chinese older adults in this study, it is a relatively new scale. Furthermore, the Cronbach's α of some dimensions (e.g., PE) of the scale are not high. Hence, more efforts are needed to examine the content validity and improve the reliability of the scale.

The fourth limitation concerns the comprehensiveness of the measure used in the survey. The present survey adopted the Social Convoy Questionnaire (Kahn & Antonucci, 1980) as the measurement tool to examine the features of social network. However, the number of question items to be included in the study has laid a constraint on the examination of social network. Therefore, some specific and detailed facets of social network (e.g., the contact frequency with social partners and the degree of perceived emotional closeness with social partners) were not covered in the survey. This limitation may result in incomplete information about different aspects of social network.

The fifth limitation is insufficient data on coping strategies. Although substantial findings related to coping in this sample have been elucidated, the process of how older people utilize coping strategies to manage stressors and the mechanism behind how coping serves as a mediator in the dynamic positive aging process remain unexamined. Specifically, more studies aiming to explore prevalent coping strategies and their effect on well-being in later life are required to fill the research gap.

10.3 Directions for Future Research

In view of the limitations present in the research design and in response to observations made in the present study, several recommendations are suggested for future research in relation to social support network, coping and positive aging.

10.3.1 Longitudinal Study

The present study used a quantitative method (i.e., survey) and a qualitative method (i.e., in-depth interview) to examine the properties of social support network, coping strategies and positive aging among Hong Kong older adults. The cross-sectional study uncovered the prevalence of peer support and companionship support among community-dwelling older adults, the mediating role of coping in the relationship between emotional closeness and overall positive aging and the pros and cons of domestic helpers in support provision. Though the findings of this study could contribute to the exploration of the social realities facing the elderly in Hong Kong, specific and well-designed longitudinal studies are highly recommended to further examine the causal relationship between the main variables. As the theoretical foundation of the present study, the convoy model of social support (Kahn & Antonucci, 1980) and SST (Carstensen et al., 1999) are characterized as a lifespan approach to explore changes in one's social network and emotional closeness across the trajectory of life. However, the present cross-sectional study limits a comprehensive consideration of the lifespan theories to explain the research findings, leaving a number of research questions unanswered. More specifically, the convoy model utilizes three different circles (i.e., the inner, middle and outer circles) to depict social network changes across one's lifespan. Similarly, SST also pertains to this feature. Although the present survey endeavored to ask the respondents to recall changes in terms of their social support network within the past ten years, the uncertainty of their answers due to loss of memory could result in unreliable findings. Hence, to acquire cumulative data, future research should be undertaken that adopts a longitudinal approach, covering different observations at different points of time across one's lifespan.

In addition, using the present cross-sectional data, the causal effect between variables could not be assessed. This begs the question, does coping only have a relationship with emotional closeness in terms of overall positive aging, or does it necessarily carry a mediating effect? Therefore, longitudinal data are needed to address these research questions.

10.3.2 Further Research in Relation to Social Support Network

Though the present study examined the structural properties (i.e., size and type) and functional properties of social support network, there are still other attributes of support network that are suggested by scholars and researchers. This study explored only perceived social support. As derived from past literature, social support network highlights the exchange of resources intended by the provider or perceived by the recipient that are considered beneficial to the recipient (Shumaker & Brownell, 1983). Consequently, social support network can be researched through two lenses: received social support versus offered social support and perceived social support versus actual social support. However, the present study only explored received social support and perceived functional social support among the elderly, failing to further address questions about the interaction within the process of support exchange. As a result, future research (e.g., a dyadic study) could be conducted to collect data from the party of support providers as well and to differentiate perceived and actual social support.

Meanwhile, the quality of social support is also an important field for further research. Antonucci, Fiori, Birditt, and Jackey (2010) highlighted the quality of social support (e.g., one's evaluation of or satisfaction with his or her social network and social support) in an updated conceptual framework using the convoy model. They indicated that individuals' evaluation of or satisfaction with their social relations would serve as an indicator together with social network (i.e., the structure of the network) and social support (e.g., exchange of aid, affect and affirmation) to get a comprehensive image of the convoys and how they experience stress. Therefore, in keeping with the theme of the updated convoy model, it is important to call for future study that looks at the quality of social support.

It is also worthwhile to propose additional research into social support network that takes a pattern-centered approach. Many researchers have explored convoys of social support using the mapping technique in social network analysis, which is classified as the variable-centered approach. Studies utilizing this approach mostly consider individual specific variables. The pattern-centered approach allows for examination of several variables at the same time, permitting the identification or description of network properties with the emphasis on the individual as part of an integrated person–environment system. However, one shortcoming of the variable-centered approach is that it fails to look upon one's social network as a dynamic system (Antonucci et al., 2010). Consequently, future research could draw support from the pattern-centered approach by using advanced analytical techniques (e.g., cluster analysis or latent class analysis) to determine network types. Through such a method, a holistic picture of one's social network pattern could be presented.

10.3.3 Further Research on Emotional Closeness

Though the present study addressed the importance of emotional closeness among community-dwelling elderly through the exploration of peer and family emotional closeness, the findings are still not conclusive. This is because the operationalization of emotional closeness in this study is not sophisticated enough. To the best of my knowledge, there is no widely recognized operationalization of emotional closeness for evidence-based research. Therefore, more research should be done to elucidate the features of emotional closeness, which could then be used to design measurement tools. Meanwhile, qualitative study into emotional closeness is recommended in order to discover more information relevant to this concept (e.g., the criteria for determining close social partners).

10.3.4 Survey of a More Representative Sample

The present survey employed a nonrandom sample of community-dwelling older Hong Kong Chinese to examine positive aging and other affecting variables. In spite of the efforts to get a diverse and representative sample of the older adults living in the community, the present survey failed to generate representativeness from the current findings because of the nonprobability sampling method. Beyond the constraint of obtaining research funding, collecting data in Hong Kong can be quite difficult. Hong Kong does not have a system where data sets are archived for public access. Hence, researchers obtain their own data sets, which are often small in scope. Collaboration between the Census and Statistics Departments is needed in order to provide access to the population sampling frame. Furthermore, respondents of the current survey were all members of the DECC; this could not represent the entire older population in the community. Hence, future surveys should use a more representative sample in order to examine issues in later life in Hong Kong as well as in other Chinese societies.

10.3.5 Validation of Expanded PAI

The 19-item expanded PAI, which was validated in the survey, consists of five components (i.e., illness avoidance, functioning, caring engagement, productive engagement and life satisfaction). The measurement of positive aging is an expanded version of the Positive Aging Index developed and validated by Ng et al. (2011). It was modified with reference to the two-factor model of successful aging (Pruchno et al., 2010). This validated tool might only pertain to the current sample, which consisted of older Chinese from community centers. Hence, future research on positive aging using a larger sample is required to further validate this tool. Furthermore, researchers could apply the framework of positive aging to other Chinese aged populations—for example, older adults living in residential care homes and the widowed elderly.

References

Antonucci, T. C., Fiori, K. L., Birditt, K., & Jackey, L. M. (2010). Convoys of social relations: Integrating life-span and life-course perspectives. In R. M. Lerner, M. E. Lamb, & A. M. Freund (Eds.), *The handbook of life-span development* (Vol. 2, pp. 434–473). Hoboken, NJ: Wiley.

Bowman, G. D., & Stern, M. (1995). Adjustment to occupational stress: The relationship of perceived control to effectiveness of coping strategies. *Journal of Counseling Psychology, 42*(3), 294–303.

Carstensen, L. L., Isaacowitz, D. M., & Charles, S. T. (1999). Taking time seriously: A theory of socioemotional selectivity. *American Psychologist, 54*(3), 165–181.

Chong, A. M. L., Ng, S. H., Woo, J., & Kwan, A. Y. H. (2006). Positive ageing: The views of middle-aged and older adults in Hong Kong. *Ageing and Society, 26*(2), 243–265.

Crowther, M. R., Parker, M. W., Achenbaum, W. A., Larimore, W. L., & Koenig, H. G. (2002). Rowe and Kahn's model of successful aging revisited: Positive spirituality—The forgotten factor. *The Gerontologist, 42*(5), 613–620.

Kahn, R. L., & Antonucci, T. C. (1980). Convoys over the life course: Attachment, roles and social support. In P. B. Baltes & O. Brim (Eds.), *Life-span development and behavior* (Vol. 4, pp. 253–286). New York: Academic Press.

Ng, S. H., Cheung, C. K. J., Chong, A. M. L., Woo, J., Kwan, A. Y. H., & Lai, S. F. Y. (2011). Aging well socially through engagement with life: Adapting Rowe and Kahn's model of successful aging to Chinese cultural context. *The International Journal of Aging and Human Development, 73*(4), 313–330.

Ng, S. H., Woo, J., Kwan, A., & Chong, A. (2006). Positive ageing: A conceptual framework. *Korean Journal of Psychological and Social Issues, 12*(5), 29–43.

Pruchno, R. A., Wilson-Genderson, M., & Cartwright, F. (2010). A two-factor model of successful aging. *The Journals of Gerontology Series B, 65*(6), 671–679.

Shumaker, S. A., & Brownell, A. (1983). A taxonomy of social support: Disentangling a conceptual morass. In A. Brownell & S. A. Shumaker (Chairs), *Annual Meeting of the American Psychological Association*, Anaheim, CA.

Social Welfare Department, Hong Kong SAR Government. (2014). *Overview of residential care services for elders.* Retrieved online from: http://www.swd.gov.hk/doc/elderly/ERCS/LTC%20statistics%20HP-Eng(201403).pdf.

Chapter 11
Conclusions

The present study examined positive aging, social support network and coping in community-dwelling older adults. It demonstrated the significance of peer support among the elderly and addressed the importance of emotional closeness in contributing to positive aging. Coping was found to serve as a mediator in the relationship between social support network and positive aging.

This section is composed of two parts. First is an examination of the study's contributions to four major areas: knowledge building, theory testing, social work practice, and policy making. And finally, the conclusions of the study are presented.

11.1 Contributions

The present study made major contributions to the following four areas: (1) knowledge building, (2) model testing, (3) policy making, and (4) social work practice.

11.1.1 Contribution to Knowledge Building

The current study attempted to contribute to knowledge building in two ways. Although the findings regarding knowledge building mainly pertain to this single study, they have significant implications. The first contribution is that this study revisited positive aging among Chinese older adults by utilizing an expanded version of PAI as its measure. Based on the conceptual framework of positive aging (Ng et al., 2011), the current survey explored the following four components: illness avoidance, functioning, CE and PE. In addition, a fifth component, life satisfaction was used. The survey adopted a two-factor model of successful aging containing objective components and subjective components (Pruchno et al., 2010). The objective components were illness avoidance, functioning, CE and PE, and the subjective

S. Liu, *Social Support Networks, Coping and Positive Aging Among the Community-Dwelling Elderly in Hong Kong*,
https://doi.org/10.1007/978-981-13-3783-3_11

component was life satisfaction. The expanded PAI (19-item) was validated as a reliable tool to explore positive aging among the community-dwelling elderly in Hong Kong. It broadens the application of measures for positive aging, revisits the two-factor model of successful aging in Chinese society, and enriches the conceptual framework of positive aging. The current survey was able to provide additional empirical evidence on positive aging and related concepts such as successful aging and productive aging in Chinese societies.

Meanwhile, it has implications regarding the effects of emotional closeness and coping strategies on positive aging and the associations among these variables. The present study also contributed the insight that peer groups could have a significant impact on positive aging in the community. This finding points out the important role of peers and the positive influence of friendships for senior citizens living in the community. In slight contrast with past empirical studies indicating that family members account for a large part of well-being in later life, the current study suggested that steady, reciprocal and meaningful relationships with peer group members (e.g., old classmates, community center members, neighbors, old colleagues and church members) could facilitate positive aging in the community. Therefore, it is recommended that future research in social gerontology further study the role of peer groups in positive aging.

11.1.2 Contribution to Model Testing

The present study tested the research hypotheses derived from the convoy model of social support. The analysis of social network composition reflected the importance of the role played by primary family members and peer groups in later life. Hypotheses testing revealed significant predictions of social network properties (e.g., size and function) and emotional closeness (e.g., family and peer emotional closeness) on positive aging. The current research lends support to a number of propositions that address the role of the family network and peer network in contributing to well-being in later life.

Additionally, the present study proposed several hypotheses related to emotional closeness (e.g., the breadth and the depth) and attempted to examine their effect on positive aging. In light of the implications of SST, emotional closeness was found to generate a crucial effect on well-being in later life. Family and peer emotional closeness, which were defined according to the depth of the relationship, were found to be positively predictive of positive aging. However, the number of close social partners, as an indicator of the breadth of emotional closeness, was not shown to be predictive of positive aging. The series of findings in the current study provide evidence and hold implications for the understanding of issues regarding emotional closeness. To the best of our knowledge, the present study is one of the very few empirical studies to examine emotional closeness and its association with well-being in later life in Chinese society. Hence, the present study has contributed a successful application of the convoy model of social support to interpret the composition of three circles of convoy and SST to explain the role of emotional closeness in contributing to well-being in Chinese societies.

11.1.3 Contribution to Policy Making

The absolute numbers of people requiring long-term care (LTC) are expected to increase as a result of a rise in life expectancy. In Hong Kong, the unusually high institutionalization rate (6.8%) among the elderly aged 65 and above points to a pressing issue faced by policy makers (Chui et al., 2009). According to Hong Kong's Secretary for Labor and Welfare, "This preference of our elders to continue living at home provides the backdrop of the Government's policy to promote aging in place (Cheung, 2009)." Additionally, Hong Kong is also on the way to construct an age-friendly society to promote the provision of suitable living environments to ensure the successful participation, health, security and independence of older adults (Hong Kong Council of Social Service, 2011).

In line with the policy directions, the findings of the study carry three important implications. First, the present study constructed a more comprehensive framework of positive aging and introduced a validated tool of positive aging—expanded PAI. It provided additional evidence for policy makers by outlining the proper pathways for service organizers to help members lead a positive later life. To facilitate the health and functioning, the society and community should build more infrastructure, facilities (e.g., outdoor seating area, flat and wide non-slip channel, green and safe crossing facilities) and caring services (e.g., day care centers, enhanced home services and community care services) to help the elderly, including frail older adults, remain in the community. Although the current sample included a large percentage of relatively healthy elders, the research holds prospects to promote AiP in a larger older population.

Second, the present study bears rich implications to one of the features of an age-friendly society—promote older adults' inclusion in and contribution to all areas of community life. The tenet of positive aging in this study advocates the elderly to maintain their social roles and remain conducive to the community and society. The findings of the study might recommend some prevalent measures to lead a productive later life, for example, participating in volunteering work, social and leisure activities, undertaking household maintenance activities, etc. Therefore, it is suggested that policy makers could formulate proper stimulative and awarding mechanism to call for more elder volunteerism. Meantime, it is advised for education scholars to extend the recognition of elder volunteerism among youngsters through education at an early age.

Third, the current research discovered the significant role of peer groups in predicting positive aging in the community, which may give inspiration to the areas of public housing (e.g., promoting elderly housing to make the older adults live in an environment with resources of peer support) and social services (e.g., proposing more projects or plans to promote neighborhood activities).

11.1.4 Contribution to Social Work Practice

In addition to the contributions to the areas of knowledge building, model testing and policy formulating, the study also revealed three insights regarding the practice of social work and social services.

First, the study showed the significant role peer support plays in the lives of elderly Chinese. As a result of these findings, more investment should be made in building a solid community network for the elderly. It is also imperative that community centers (e.g., DECC, NEC and SCE) appeal to older adults and organize more social, sports, recreational, volunteering and leisure activities for their members. It should be noted that although a number of older members have registered, they seldom participate in organized activities or make regular contact with the social workers or other center members (Social Welfare Department, 2006). Social workers should stay in touch with center members who seldom show up and put effort into meeting their needs, both physical and emotional. It is also recommended that social workers offer opportunities for elderly individuals to have meaningful interactions within their peer group. It has been recognized through the practical experience of the researcher that home visits by older volunteers would likely improve the well-being of the widowed or elders who live alone. These individuals are likely to be uplifted through communicating with visitors from other age groups (e.g., children and students), with whom they can access optimism and better emotional states.

Second, it is recommended that social workers be attentive to the caregiving needs of the elderly who could benefit from additional caregiving. The social workers could talk with the elders or their family members to help them identify the type of domestic help they need, whether it be FDH, part-time help, enhanced home service or community care service. In addition, social workers could help solve the conflicts between the elderly, primary family members and DHs, and offer appropriate tips to mediate their interrelationships.

Third, regarding counseling older people, social workers are advised to encourage the elderly to practice more caring engagement and productive engagement if possible. Social workers could help their older clients understand positive aging and encourage them, if capable, to take part in caring engagement, such as household maintenance, emotional or financial support for younger generations, volunteerism, and part-time work. In addition, employers could be encouraged to create more part-time jobs or appoint capable retired staff members as advisors in their field of expertise; this would cultivate productive engagement in later life.

11.2 Conclusion

The present research consisted of two phases of study. The pilot study was a qualitative study that employed in-depth interviews with 16 older participants from DECCs of two districts in Hong Kong. Interview guidelines were constructed and thematic analysis was conducted for data analysis. The main study was a cross-sectional survey of a total of 393 older people who were selected from six DECCs of five districts in Hong Kong through a multistage sampling method. To measure the dependent variable—positive aging—the study adopted Ng et al.'s tool, the Positive Aging Index (PAI), and expanded it with the implications from a two-factor model of successful aging (Pruchno et al., 2010). The questionnaire used in the survey was constructed and revised according to the important findings from the pilot study. Data analysis was conducted by virtue of IBM SPSS 21.0 in conjunction with Amos 20.0, by which bivariate analysis, stepwise multiple regression, mediation test and path analysis were performed.

This study is one of the very few empirical studies to explore social support network, coping and positive aging among community-dwelling elderly in Hong Kong. It expanded the original concept and measurement of positive aging into a two-factor construct and validated the expanded Positive Aging Index comprising illness avoidance, functioning, caring engagement, productive engagement and life satisfaction. It identified the structural properties (i.e., size and type) and functional properties (e.g., emotional support, companionship support, instrumental support and informational support) by adopting the convoy model of social support. In addition, this study examined the interrelationships between social relations (e.g., emotional closeness), coping strategies (e.g., problem coping, emotion coping and avoidance coping) and overall positive aging, confirming the mediating effect of coping. The study supported the contribution of social network (e.g., social network size and family and peer networks), family and peer emotional closeness and coping strategies (e.g., problem and emotion coping) to overall positive aging. Further, it described the types of caregiving services provided by domestic helpers and the advantages and disadvantages of domestic helpers in looking after community-dwelling elderly in Hong Kong.

The present study depicted an indigenous landscape of positive aging of older Hong Kong Chinese in the community center from lower-middle class and clearly presented the contributing factors including social network size, family and peer emotional closeness and coping strategies. These significant factors were deemed essential to enable the Hong Kong elderly to age well in the community, shedding light on the meaning and benefits of the notion of *aging in place*. These findings have implications for communities and families wishing to assist the elderly in leading a positive life.

The present study offered a variety of recommendations from the domains of policy making, social work practice, elderly services and research. Promoting positive aging in Chinese societies, consolidating the social support network and improving the elders' coping capacity are outlined for future action.

The limitations of this study have been duly noted. In addition, avenues for future research that could answer the remaining research questions in a lifespan perspective were identified. The results and implications of the current research regarding the concept, conceptual framework and measurement of positive aging require further demonstration and extension in other samples at home and abroad, motivated by the desire to promote well-being in later life.

References

Cheung, M. (2009). Government promotes ageing in place. Address given at the '*CADENZA Symposium 2009: Primary Care & Older Persons—Key to Medical & Social Integration*. Retrieved online from http://www.info.gov.hk/gia/general/200910/09/P200910090092.htm.

Chui, E. W. T., Ko, S. F. L., Law, C. K. S., Lou, W. Q. V., Ng, Y. T. S., Chan, K. S., …, Leung, Y. M. A. (2009). *Elderly commission's study on residential care services for the elderly final report*. Hong Kong, China: Elderly Commission.

Hong Kong Council of Social Service. (2011). *The eight domains in relation to "an age-friendly city", the age-friendly society project*. Hong Kong Council of Social Service.

Ng, S. H., Cheung, C. K. J., Chong, A. M. L., Woo, J., Kwan, A. Y. H., & Lai, S. F. Y. (2011). Aging well socially through engagement with life: Adapting Rowe and Kahn's model of successful aging to Chinese cultural context. *The International Journal of Aging and Human Development, 73*(4), 313–330.

Pruchno, R. A., Wilson-Genderson, M., & Cartwright, F. (2010). A two-factor model of successful aging. *The Journals of Gerontology Series B*, *65*(6), 671–679.

Social Welfare Department. (2006). *Executive summary of DECC and NEC*. Retrieved online from www.swd.gov.hk/doc/…/ExecutiveSummary_chi.doc.

Appendix A

In-depth interview guidelines (English version)

Dear friends:

Hello! I'm a Ph.D. student of Dept. of Applied Social Sciences, City University of Hong Kong. Now I'm conducting a research on the social support network, coping and positive aging among the older adults. At present I'm inviting you to participate in this interview which generally takes 30–45 min. During the interview process, a serious of questions related to the research topic will be asked. You are welcome to answer them according to your personal experience.

In order to process data, the interview process will be taped. All the information and materials will be strictly secured and be used in my doctoral thesis. Additionally, the researcher will not relate your answers to your personal information.

A small gift will be presented to you by way of appreciation.

Wish you good health and happy life.

Consent Form

- ☐ I have known about the research objectives and process.
- ☐ I have known that all the personal information will be confidentially secured.
- ☐ I consent to participate in this study and am aware that I can quit at any time during the study.

Signature _________ Year ___ Month ___ Date __

S. Liu, *Social Support Networks, Coping and Positive Aging Among the Community-Dwelling Elderly in Hong Kong*,
https://doi.org/10.1007/978-981-13-3783-3

No.: ___
Interview Time: *2012* __ Month __Date

Part 1: Personal Information

Name: ______
Age: ___
District: ____
Marital status: ____
No. of children: ___

Part 2: Social Support Network

1. How long have you stayed in Hong Kong? Who do you live with?
2. In your memory, when did you feel you had stepped into old age? Or what event led to old age?
3. What are the main challenges after you step into old life? Which challenge is the most difficult to overcome?
4. What about your current health status? Are you self-care capable?
5. Have you or your family members employed domestic helpers? Foreign domestic helper or local? Full-time or part-time? What kind of services does she provide?
6. If you encounter some difficulties in life, who would you like to turn to? Please nominate them.
7. If you come across unhappy events, would you like to find someone to talk to?
8. Who do you usually eat or chat with?
9. Compared with younger times, please recall if there is any change in terms of the social support, help and caring you receive at present?

Part 3: Coping

1. What types of stress did you come across in the process of getting old? How did you cope with them?
2. (Only applicable to the participants who are self-care incapable) As mentioned above, you need assistance in your daily life, do you fell unhappy or stressful? If so, what strategies did you employ to get happier? Did you try any methods to become healthier?
3. Have you ever tried to change the situation that you are receiving less social support or experiencing maladjustment in life and mental state?

Part 4: Positive Aging

1. Recently, have you helped caring your family (e.g. undertaking housework, fostering grandchildren and cooking)? Are you willing to do these?
2. Do you work now? Part-time or full-time? Why do you work now?
3. Do you participate in the below activities regularly? For example, volunteering, training class, interest group, voting, social activities, artistic activities, sports and political activities?
4. In general, are you satisfied with your current life? What are the satisfactory facets and what are the unsatisfactory facets?

In-depth interview guidelines (Chinese version)

尊敬的長者：

你好。我是香港城市大學應用社會科學系的一名博士研究生，現在正開展一項有關老年人社會支持網路(個人從社會中獲得的支持、關懷與幫助等）與黃金歲月的研究。現邀請你參與一個訪談，訪談時間大概 30—45 分鐘。訪談過程中我會問一些與研究相關的問題，希望閣下可以根據自身情況回答。

為方便數據處理，訪談過程將被錄音。本研究所獲得的所有資料將完全用於我的博士論文研究，你所提供的任何資料及信息都是絕對安全及保密的，同時，亦不會將你的任何回答聯繫到你的個人資訊。

為表謝意，訪問結束后，我會送贈紀念品作為回饋予閣下。

祝你身體健康，生活開心。

<u>同意書</u>

本人已詳知有關這項研究的目的及過程。

本人明白所有關於本人的個人資料均絕對保密。

本人明白訪談的過程將被錄音并只會用于研究用途。

本人同意參與這項研究，并了解到本人可于研究進行期間有權隨時退出。

簽署_________ 日期 2012 年___月___日

編號：___
訪談時間：2012 年__ 月 __日

第一部份：基本情況

姓名：______
年齡：___
居住地：____
婚姻狀況：____
子女情況：___
有無家庭傭工：____

第二部分：社會支持網路

1. 你在香港居住有多少年了？與誰一起住？

2. 在你的記憶中，大約從什麼時候開始，或者什麼事件讓你感覺自己已經步入老年？

3. 在步入老年之後，你認為主要面臨的挑戰有哪些？在這些挑戰中，你認為最難克服的是哪一點？

4. 你目前的身體狀況如何？可以自理嗎？

5. 你或你的家庭成員現在或曾經有請過傭人嗎？是外傭還是本地傭人？（全職還是兼職？她平時主要幫助你或你的家人做些什麼工作？）

6. 如果你在現實生活中遇到困難，你願意向誰尋求幫助？請列舉幾例。

7.如果你遇到煩惱或不開心的事情，你會不會找人傾訴？

8. 平時你經常和誰在一起吃飯、聊天？

9. 請回顧一下，與年輕的時候相比，別人對你的支持、幫助及關懷有沒有變化？

第三部分：應對方式

1. 在年齡逐漸增長的過程中，你經常會遇到哪些壓力？你是如何應對這些壓力的？

2.（只對身體狀況不好的被訪者適用）從剛才的談話中可以了解， 目前你的生活不能自理，仍然需要別人的幫助，你是否會感到不開心或為此而煩惱？（如果是，平常你都會採取什麼方式讓自己變得更開心？有沒有採取什麼方法試圖讓身體更健康一點？）

3. 對于一些生活、心理上的不適應，或是其他人提供的支持過少，你有沒有自己獨特的方式去調節或是去改變？

第四部份：黃金歲月

1. 最近或目前，你是否有幫到屋企？（如承擔一些家務工作，照顧孫兒，燒飯等。）你是否願意为家庭做這些事情？

2. 你現在有工作嗎？是全職還是兼職的？你現在為什麼要工作？有何感受？

3. 你平時有參加什麼活動？（如義務工作，學習，培訓課程，興趣小組，投票等政治性活動）

5. 總的來說，你對你目前的生活狀態感到滿意嗎？滿意之處有哪些方面？不滿意之處有哪些方面？

Appendix B

Survey questionnaire (English version)
****

No.: □□□

Dear Center Members:

Hi! I'm the interviewer of this study. My name's _________ from City University of Hong Kong. At present, a survey is being conducted by Miss Liu from the Dept. of Applied Social Sciences. The survey is to explore the social support network, coping and positive aging among the community-dwelling older adults in Hong Kong. I sincerely invite you to participate in this survey. The interview process takes around 30–45 min. You are invited to answer the questions raised in the questionnaire, which are related to the topic.

All the information collected will be entirely used in this study. Your personal information and materials provided will be strictly secured. In the meantime, your answers would not be deliberately related to your personal information.

You will be presented a small gift by way of appreciation.

Wish you good health and happy life.

Consent Form

□ **I have known about the research objectives and process.**
□ **I have known that all the personal information will be confidentially secured.**
□ **I consent to participate in this study and am aware that I can quit at any time during the study.**

Signature _________ Year ___ Month ___ Date __

S. Liu, *Social Support Networks, Coping and Positive Aging Among the Community-Dwelling Elderly in Hong Kong*,
https://doi.org/10.1007/978-981-13-3783-3

Interview Records:

District: _________ Institute: _________
Contact Person: _________
Interviewee name or solution: _________

Times	**Dates**	**Starting time**	**Ending time**	**Completed or not**	**Partially incompleted**	**Refusal**	**Others**
1	Year___ Month___Date__ _	--- Hour --- Min	-- Hour -- - Min				
2	Year___ Month___Date__ _	--- Hour --- Min	-- Hour -- - Min				
3	Year___ Month___Date__ _	--- Hour --- Min	-- Hour -- - Min				

Interviewer Signature : ________________ Date : 201__Year____Month____Date

Entry Staff Signature : ________________ Date :2013 Year_____Month_____Date

Part 1 Screening Test

1. Gender： (Not ask but observation) (1) Male (2)Female
2. Education Background：
(1) Illiterate (2) Old-style private school (3) Primary school (4) Secondary school (Grade 1 to 5) (5) Foundation (Grade 6 to 7) (6)Post-secondary education (Diploma, Certificate, Associate degree, Degree and Postgraduate.)
3. We will ask you a few simple questions to test your memory. These questions could not be memorized by everyone. Please try to recall.

Questions	Guidance
(1) What day is today?	
(2) What's the date today?	
(3) Who is the newly-elected EO?	C.Y. Leung
(4) Who is the last EO?	Donald Tsang
(5) Can you count backward from 20 by 3's?	Take *go to market and buy food* as example
(6) a What is your phone number? (If no, then ask (6)b)	
(6) b What's your address (e.g.district, road or building) ?	
(7) What's the district of this DECC (HKI, KL or NT) ?	
(8) How old are you ? _______	Check with the ID (if any)
(9) When were you born ? ---Y---M----D	Check with the ID (if any)
(10) If illiterate or old-style private school, add 1 point.	--------------------
Total	--------------------

【Notes】3points or below 3 points. Suspend the interview and thanks for the cooperation.
4 points or above 4 points. The interview continues.

Part 2 Family Support Network

4. How many relatives do you see or hear from at least once a month (Considering the people to whom you are related by birth, marriage, adoption, etc. …)?
(1) 0； (2) 1； (3) 2； (4) 3 or 4 ； (5) 5 thru 8； (6) 9 or more
5. How often do you see or hear from the relative with whom you have the most contact?
(1) less than monthly； (2) monthly ； (3) few times a month
(4) weekly ； (5) few times a week； (6) daily
6. How many relatives do you feel at ease with that you can talk about private matters?
(1) 0； (2) 1 ； (3) 2 ； (4) 3 or 4； (5) 5 thru 8； (6) 9 or more
7. How many relatives do you feel close to such that you could call on them for help?
(1) 0； (2) 1 ； (3) 2 ； (4) 3 or 4； (5) 5 thru 8 ； (6) 9 or more
8. When one of your relatives has an important decision to make, how often do they talk to you about it?
(1) never； (2) seldom； (3) sometimes； (4) often； (5) very often； (6) always
9. How often is one of your relatives available for you to talk to when you have an important decision to make?
(1) never； (2) seldom； (3) sometimes； (4) often； (5) very often； (6) always
10. Compared with ten years ago, the family support you have received at present is :
(1) less than before； (2) almost the same； (3) more than before

Part 3 Peer Support Network

11. How many of your friends do you see or hear from at least once a month? (Considering the friends, neighbors, old classmates, church members, center members, old colleagues, etc.)
(1) 0； (2) 1 ； (3) 2 ； (4) 3 or 4； (5) 5 thru 8 ； (6) 9 or more
12. How often do you see or hear from the friend with whom you have the most contact?
(1) less than monthly； (2) monthly； (3) few times a month；
(4) weekly； (5) few times a week； (6) daily
13. How many friends do you feel at ease with such that you can talk about private matters?
(1) 0； (2) 1 ； (3) 2 ； (4) 3 or 4； (5) 5 thru 8 ； (6) 9 or more
14. How many friends do you feel close to such that you could call on them for help?
(1) 0； (2) 1 ； (3) 2 ； (4) 3 or 4； (5) 5 thru 8 ； (6) 9 or more
15. When one of your friends has an important decision to make, how often do they talk to you about it?
(1) never； (2) seldom； (3) sometimes； (4) often； (5) very often； (6) always
16. How often is one of your friends available for you to talk to when you have an important decision to make?
(1) never； (2) seldom； (3) sometimes； (4) often； (5) very often； (6) always
17. Compared with ten years ago, the peer support you have received at present is
(1) less than before； (2) almost the same； (3) more than before

Part 4 Social Network Composition

18. Please recall throughout your life and consider that there are some persons who are very important to you, some others are less important.

(1).the inner circle (social partners you feel so close with that you cannot image life without them):

(2).the middle circle (social partners who are import but rather close ones):

(3).the outer circle(social partners who are less close and less important):

【Notes】 Interviewers write down the names first and classify the answers into the below categories, and then calculate the number of each category. (If there is no nomination, fill 0.)

A.primary family members (e.g. parents, spouse, adult children, grandchildren and cousins)

B.peer group(e.g. friends, coworkers, elderly community centre members, church members, neighbors, classmates, etc.)

C.domestic helper (e.g. live-in foreign domestic helper, part-time or full-time local helper)

D.relatives (e.g. relatives other than primary family members)

E.professionals (e.g. doctors, nurses, social workers, physicians, etc.)

(1)Total____ A___ B___ C___ D___ E____
(2)Total____ A___ B___ C___ D___ E____
(3)Total____ A___ B___ C___ D___ E____

Part 5 Functional Support Network

Please recall when you are in trouble or need help, do the persons nearby can provide support to you? Please tick the most proper answer according to your own experience.

19. (a) Please rate the below relationships according to the statement "listening to you, reassuring you, showing you that they care and are willing to share happiness and sorrows".

	① never	② seldom	③ sometimes	④ often	⑤ very often
(1) Primary family members	□	□	□	□	□
(2) Peer group	□	□	□	□	□
(3) Domestic helper	□	□	□	□	□
(4) Relatives	□	□	□	□	□
(5) Professionals	□	□	□	□	□

(b) Compared with ten years ago, the emotional support you have received at present is
(1) less than before; (2) almost the same; (3) more than before

20.(a) Please rate the below relationships according to the statement "how often during the previous years do you receive support with daily chores around the house (e.g. preparing meals, cleaning the house, transportation, and small repairs, etc.) and caregiving"

	① never	② seldom	③ sometimes	④ often	⑤ very often
(1) Primary family members	□	□	□	□	□
(2) Peer group	□	□	□	□	□
(3) Domestic helper	□	□	□	□	□
(4) Relatives	□	□	□	□	□
(5) Professionals	□	□	□	□	□

(b) Compared with ten years ago, the instrumental support you have received at present is
(1) less than before; (2) almost the same; (3) more than before

21. (a) Please rate the below relationships according to the statement "how helpful is each type of network in spending time with you in leisure and recreational activities, in offering a sense of friendship, and in participating in shared activities".

	① never	② seldom	③ sometimes	④ often	⑤ very often
(1) Primary family members	□	□	□	□	□
(2) Peer group	□	□	□	□	□
(3) Domestic helper	□	□	□	□	□
(4) Relatives	□	□	□	□	□
(5) Professionals	□	□	□	□	□

(b) Compared with ten years ago, the instrumental support you have received at present is
(1) less than before ; (2) almost the same ; (3) more than before

22. (a) Please rate the below relationships according to the statement "how helpful is each type of network in offering you useful information in terms of medicine, health, insurance, social welfare, social activities, daily news and occupation".

	① never	② seldo	③ sometimes	④ often	⑤ very often
(1) Primary family members	□	□	□	□	□
(2) Peer group	□	□	□	□	□
(3) Domestic helper	□	□	□	□	□
(4) Relatives	□	□	□	□	□
(5) Professionals	□	□	□	□	□

(b) Compared with ten years ago, the instrumental support you have received at present is
(1) less than before ; (2) almost the same ; (3) more than before

Part 6 Positive Aging

I Illness Avoidance

Over the past month

23. Has any bodily pain affected your performance of work or other essential tasks?
(1) never ; (2) rarely ; (3) occasionally ; (4) sometimes ; (5) often

24. Have you used medication or therapy to maintain regular daily life (high pressure, diabetes are excluded)?
(1) never ; (2) rarely ; (3) occasionally ; (4) sometimes ; (5) often

25. Were you healthy enough to move around freely?
(1) very poor ; (2) poor ; (3) normal ; (4) good ; (5) very good

26. Overall, your present health is
(1) very poor ; (2) poor ; (3) normal ; (4) good ; (5) very good

II Functioning

27. Recently, do you have enough energy for daily life?

(1) never; (2) rarely; (3) occasionally; (4) sometimes; (5) often

28. Have you been sleeping well recently?

(1) very poor; (2) poor; (3) normal; (4) good; (5) very good

29. In the past month, when you tried to recall familiar names or words, how often were you unable to do so?

(1) never; (2) once or twice; (3) 3-4 times; (4) 5-6 times; (5) more than 7 times

30. Can you perform two or more tasks simultaneously, for example, watch TV while discussing something else with another person?

(1) not capable; (2) slightly capable; (3) normal; (4) capable; (5) very capable

31. Presently, are your body and mind strong enough to enable you to live independently without having others to take care of you?

(1) not capable; (2) slightly capable; (3) normal; (4) capable; (5) very capable

III Caring Engagement

32. With respect to the below behaviors, how much concern and support have you provided to enrich the lives of your family members and relatives in the past year?

	① very much	② much	③ normal	④ little	⑤ very little
(1) Undertaking household maintenance activities (e.g. cooking, washing, cleaning)	□	□	□	□	□
(2) Showing care and support to close family members (i.e. spouse, adult children and grandchildren)	□	□	□	□	□
(3) Showing care and support to relatives (siblings, kin)	□	□	□	□	□

IV Productive Aging

33. How often have you participated in the below activities and make contributions to the community and society in the past year?

	① very much	② much	③ normal	④ little	⑤ very little
(1) Voluntary work	□	□	□	□	□
(2) Paid work (e.g. part-time or full-time)	□	□	□	□	□
(3) Social and leisure activities (i.e. social activities, recreational, artistic and spiritual activities, lifelong learning, civic engagement)	□	□	□	□	□

V Life Satisfaction

34. Do you agree with the current life? Please indicate your agreement with each item.

	① strongly disagree	② disagree	③ neither agree nor disagree	④ agree	⑤ strongly agree
(1) In most ways my life is close to my ideal	□	□	□	□	□
(2) The conditions of my life are excellent	□	□	□	□	□
(3) I am satisfied with life	□	□	□	□	□
(4) So far I have gotten the important things I want in life	□	□	□	□	□
(5) If I could live my life over, I would change almost nothing	□	□	□	□	□

Part 7 Coping

35. Please rate the frequency of adopting the below ways of coping when you are confronted with stress, daily hassles and problems.

	① never	② sometimes	③ often
(1) study or work	□	□	□
(2) talk with others	□	□	□
(3) see the bright side of things	□	□	□
(4) change one's thinking	□	□	□
(5) do not take things too seriously	□	□	□
(6) stand your ground	□	□	□
(7) find several ways out	□	□	□
(8) seek advices from relatives and friends	□	□	□
(9) change behaviors	□	□	□
(10) learn from others	□	□	□
(11) actively participate in recreational and sports activities	□	□	□
(12) emotion control	□	□	□
(13) have a rest or holiday	□	□	□
(14) smoke, drink, take medicine and eat	□	□	□
(15) wait	□	□	□
(16) try to forget	□	□	□
(17) rely on others	□	□	□
(18) accept the reality	□	□	□
(19) imagine miracles	□	□	□
(20) comfort yourself	□	□	□

Part 8 Personal Information

36. What is your marital status ?
(1) never married; (2) separated; (3) married; (4) divorced; (5) widowed ; (6) others (please indicate) __________ 。

37. What is your current housing type :
(1) public housing ; (2) elder residence ; (3) self-owned housing ;
(4) renting the floor ; (5) renting an apartment ; (6) renting a bed ;
(7) penthouse/wooden house ; (8) joint venture with government (9) other residence (please indicate) :

38. Household members and cohabitant (Who are you living with in an apartment or unit) ?

	No.	No. of persons who live together
Live alone		
Spouse		
Parents (alive)		
Son		
Daughter		
Daughter-in-law		
Son-in-law		
Grandchildren		
Brothers and sisters		
Domestic helper		
Friends		
Relatives		

Introduction : If the respondent lives alone, then fill 1 in the blank for live alone; if the respondent lives with others, then fill 0; fill the no. of persons with whom they live together(for instance, if the respondent lives with three sons, then fill 3 in the blank for son)

39. What is your current monthly income ? (e.g. : income, pension, CSSA, old age allowance, living expenses provided by children.) (HKD)
(1) none ; (2) below 2000 ; (3) 2001-4000 ; (4) 4001-6000 ;
(5) 6001-8000 ; (6) 8001-10000 ; (7) above 10000 ; (8) not clear

40. **What is your household's monthly income ?** (including primary family members living in the same roof) (HKD)
(1) below 5000 ; (2) 5001-10000 ; (3) 10001-15000 ; (4) 15001-20000 ;
(5) 20001-25000 ; (6) 25001-30000 ; (7) 30000 and above 。

41. Generally, how do you rate the socioeconomic status of your household in Hong Kong?
(1) lower ; (2) lower-middle ; (3) middle ; (4) upper-middle ; (5) upper

42. Do you think you have enough money to spend?
(1) quite enough; (2) enough; (3) a little enough; (4) not enough; (5) far from enough

43. What was your occupation before retirement?
(1) ordinary worker (e.g. waiter, sales, gatekeeper, construction worker) ;
(2) skilled worker (e.g. driver, maintenance worker)
(3) task master (e.g. construction headman)
(4) self-managed business (e.g. taxi driver, grocery store boss)
(5) civilian post (e.g. secretary, agent, research assistant)
(6) technician (e.g. computer officer, social worker, teacher, nurse)
(7) professional (e.g. lawyer, engineer, professor and headmaster)
(8) administration staff (e.g. manager, director)
(9) policeman
(10) CEO
(11) housewife or househusband
(12) others (please indicate : _______________)

44. Are you suffering with the below illness? (you can choose more than one)
(1) none ; (2) high/low blood pressure ; (3) eye diseases, e.g. cataract ; (4) tracheopathy ; (5) heart disease ; (6) arthritis (foot ache, body ache, shoulder ache) ; (7) diabetes; (8) stroke; (9) nephroma; (10)

parkinson; (11) dementia (12) arthrolithiasis (13) others(please indicate: ------------------)

45. Do you need assistance in daily life? If so, who is your primary caregiver?

(1) spouse; (2) son; (3) daughter; (4) daughter-in-law; (5) son-in-law; (6) grandchildren; (7) domestic helper; (8) relatives; (9) friends; (10) others(please indicate: --------------)

46. Do you employ domestic helper now? (If choose 1, then skip 52-56)

(1) no; (2) full-time foreign domestic helper; (3) full-time local helper; (4) part-time (hourly employee)

47. What kind of services do domestic helper provide? (can choose more than one)

(1) basic housework (cooking, cleaning); (2) personal care (bathing, cutting fingernails, dressing); (3) assisting in medical care and nursing; (4) looking after children and family members; (5) escorting (going out or see a doctor); (6) chatting; (7) others (please indicate:)

48. Are you satisfied with the services provided by domestic helpers?

(1) extremely not satisfied; (2) not satisfied; (3) neutral; (4) satisfied; (5) extremely satisfied

49. How much do you need to pay the domestic helper? [______] HKD

50. Who pays for the salary of domestic helper? (can choose more than one)

(1) oneself; (2) spouse; (3) son; (4) daughter-in-law; (5) daughter; (6) son-in-law;

(7) grandchildren; (8) relatives; (9) friends; (10) others(please indicate __________)

51. If someone share the salary of the domestic helper with you, how much do you need to pay monthly __________ ?

52. Are you willing to employ a domestic helper to look after you if your health deteriorates a couple of years later?

(1) yes (2) no (3) not clear

53. Can you talk about the pros and cons of employing domestic helpers as caregivers at home?

----

54. If needed, are you willing to join the Enhance Home and Community Care Services? For example, professionals will provide caregiving, recovery services and basic care, etc. to reach the goal of aging in place.

(1) Yes (2) No (3) Not clear

55. If your health condition deteriorates or becomes older, are you willing to move to residential care homes?

(1) Yes (2) No (3) Not clear

56. Are you religious?

(1) none; (2) Christian; (3) Catholicism; (4) Buddhism; (5) ancestor worship; (6) Islam; (7) Taoism; (8) others (please indicate_________________)

This is the end, thank you for your cooperation!
****

Notes of interviewers

Does the interviewee have serious mental problems? Does he or she need referral to social workers?	
Did you encounter any problems or obstacles during the process?	
What did you learn from the interview?	

Survey questionnaire (Chinese version)

****

編號：□□□

尊敬的長者：

你好。我是本次研究的訪問員---------------，是香港城市大學的學生。目前，應用社會科學系正開展一項由博士研究生劉素素小姐發起的研究，探索香港老年人的社會支持網路（個人從社會中獲得的支持、關懷與幫助等）與如何度過黃金歲月。現邀請你參與一個问卷调查，訪談時間大概九個字。訪談過程中會問一些與研究相關的問題，希望閣下可以根據自己情況回答。

本调查所獲得的所有資料將完全用於研究，你所提供的任何資料都是絕對安全及保密的，同時，亦不會將你的任何回答聯繫到你的個人資料。

以表謝意，訪問結束后，研究者會送贈紀念品。

祝你身體健康，生活開心。

<u>同意書</u>

□本人已詳知有關這項研究的目的及過程。
□本人明白所有關於本人的個人資料均絕對保密。
□本人同意參與這項研究，並了解到本人可于研究進行期間有權隨時退出。

簽署__________ 日期 2012 年___月___日

訪問記錄：

區：-------------------　　機構：----------------------
中心聯絡人：----------------
老年人姓名或稱謂：---------------------

問次數	訪問日期	開始訪問時間（24小時制）	結束訪問時間（24小時制）	成	部份未完成	拒絕訪問	其他
	20___ 年 ___ 月 ___ 日	--- 時 --- 分	--- 時 --- 分				
	20___ 年 ___ 月 ___ 日	--- 時 --- 分	--- 時 --- 分				
	20___ 年 ___ 月 ___ 日	--- 時 --- 分	--- 時 --- 分				

访问员签名：______________________　　日期：2012 年_____月_____日

录入员签名：______________________　　日期：2013 年_____月_____日

第一部分： 記憶測試

1. 受訪者性別：（訪問員觀察而不用問）　(1) 男　(2)女
2. 文化程度：
(1) 不認字　(2) 私塾　(3) 小學　(4)中學(中 1 至中 5)　(5) 預科（中 6 至中 7）　(6)專上教育（證書課程，文憑課程，副學士，學士，碩士及以上）
3. 以下我們將問你一些簡單的問題，了解一下你的記憶力，這些問題不是每個人都能夠記得的，請盡量回答。

	指導語
(1) 今天是星期幾？	星期對才算正確。
(2) 今天是多少號?	年、月、日都對才算正確。
(3) 誰是新當選的特首？	梁振英
(4) 誰是上一任特首？	曾蔭權
(5) 20 減去 3，再減去 3，一直減下去，減 5 次，答案是多少？	可用去街市買菜的例子
(6) a 你的電話號碼是多少號？(如無電話，就問 (6)b)	
(6) b 你的地址是哪兒（如：哪個社區，哪條街道，哪棟樓）？	
(7) 現在的老人中心是屬於哪一區（香港、九龍或新界）？	
(8) 你有多大年紀？------歲？	須與身份證核對（如果有）
(9) 你何時出生的？---年---月----日	須與身份證核對（如果有）
	孫文或孫中山
	---------------------
(10) 如果文化程度為“不認字”或“私塾”，則加 1 分	
總分	---------------------

【計分說明】3 分或 3 分以下，感謝長者的配合，并結束訪問。
4 分或 4 分以上，餘下問題可以由被訪者自行回答。

第二部分 家庭支持網絡

4. 有多少親人你最少一個月見面或交談一次？（注：包括父母，配偶，子女及孫子女，兄弟姐妹及其他親戚）
⑴ 0 位； ⑵ 1 位； ⑶ 2 位； ⑷ 3 或 4 位 ； ⑸ 5 至 8 位； ⑹ 9 位或更多 。

5. 請告訴我你最常接觸的親人，你與他們見面或交談多少次？
⑴ 每月少過一次； ⑵ 每月一次 ； ⑶ 每月數次
⑷ 每個星期一次 ； ⑸ 每星期數次； ⑹ 每天都有。

6. 你身邊有多少位親人能讓你訴說心事？
⑴ 0 位； ⑵ 1 位； ⑶ 2 位 ； ⑷ 3 或 4 位； ⑸ 5 至 8 位； ⑹ 9 位或更多。

7. 有多少親人讓你感到能夠親近，可以向他們尋求幫助？
⑴ 0 位； ⑵ 1 位； ⑶ 2 位 ； ⑷ 3 或 4 位； ⑸ 5 至 8 位； ⑹ 9 位或更多。

8. 如果你的親人將要做一個重要的決定，他們會跟你商量嗎？
⑴ 從無； ⑵ 很少； ⑶ 間中； ⑷ 好多次； ⑸ 幾乎每一次； ⑹ 每一次都商量。

9. 當你有重要的事情要決定時，你可以找到親人商量嗎？
⑴ 從無； ⑵ 很少； ⑶ 間中； ⑷ 好多次； ⑸ 幾乎每一次； ⑹ 每一次都搵得到。

10. 和過去（十年前）相比，現在親人給你的支持：
⑴ 比過去少； ⑵ 差不多或基本相同； ⑶ 比過去多。

第三部分 朋友支持網絡

11. 你有多少朋友一個月見面或交談一次？（注：包括同事，工友，朋友，舊同學，中心會員，教友）
⑴ 0 位； ⑵ 1 位； ⑶ 2 位 ； ⑷ 3 或 4 位； ⑸ 5 至 8 位； ⑹ 9 位或更多。

12. 請告訴我關於你最常接觸的朋友，你與他/他們見面或交談的次數是：
⑴ 每月少過一次； ⑵ 每月一次； ⑶ 每月數次；
⑷ 每個星期一次； ⑸ 每星期數次； ⑹ 每天都有。

13. 你身邊有多少位朋友能讓你訴說心事？
⑴ 0 位； ⑵ 1 位； ⑶ 2 位 ； ⑷ 3 或 4 位； ⑸ 5 至 8 位； ⑹ 9 位或更多。

14. 你有多少位朋友可以讓你感到親近，可以向他們求助？
⑴ 0 位； ⑵ 1 位； ⑶ 2 位 ； ⑷ 3 或 4 位； ⑸ 5 至 8 位； ⑹ 9 位或更多。

15. 如果你的朋友將要做一個重要的決定，他們會跟你商量嗎？
⑴ 從無； ⑵ 很少； ⑶ 間中； ⑷ 好多次； ⑸ 幾乎每一次； ⑹ 每一次都商量。

16. 當你有重要的事情要決定時，你可以找到朋友商量嗎？
⑴ 從無； ⑵ 很少； ⑶ 間中； ⑷ 好多次； ⑸ 幾乎每一次； ⑹ 每一次都搵得到。

17. 和過去（十年前）相比，現在朋友給你的支持：
⑴ 比過去少； ⑵ 差不多或基本相同； ⑶ 比過去多。

第四部分 社會網絡構成

18. 請回憶一下，在你的生活中，有一些人對你而言是十分重要的，有一些人的重要程度則會弱一些。

(1). 請你思考一下，在你的生活中，有哪些人是你認為**最重要的而且不可取代的**？一共有多少個非常重要的人，他們是誰？

(2). 請你思考一下，在你的生活中，有哪些人是你認為**比較重要**的人，一共有多少個比較重要的人，他們是誰？

(3). 請你思考一下，在你的生活中，有哪些人是你認為**不是很重要、但是卻與他們有聯絡**的人，一共有多少個這樣的人，他們是誰？

【說明】訪問員可以先記錄下答案，然後根據下面的分類將答案歸類，并計算出每一個類別的數目，最後計算總數。(若無，則填 0)

A.主要家庭成員（注：包括父母、配偶、子女、孫仔孫女、兄弟姐妹）
B.朋友 （注：包括同事、工友、朋友、舊同學、中心會員、教友或鄰居等）
C.家庭傭工 （注：包括外傭、本地的，全職或兼職的）
D.親戚 （注：包括主要家庭成員以外的親屬關係）
E.社會機構與專業人士 （注： 包括福利機構、社區中心、老人院，醫生、護士、社工等）
(1)總計____ A___ B___ C___ D___ E____
(2)總計____ A___ B___ C___ D___ E____
(3)總計____ A___ B___ C___ D___ E____

第五部分 功能性社會支持網絡

請回憶一下，在你有困難需要幫助的時候，你身邊的人是否經常提供以下的支持給你？請根據他們提供 的支持 情況，進行選擇。

19. (a) 在我遇到困難或是傷心難過的時候，他（她）能夠“從情緒上支持我，耐心傾聽，或者安慰我，或者關心我，或者願意與我分享開心與痛苦的事”

	①從無	②很少	③間中	④好多次	⑤幾乎每一次
(1) 主要家庭成員	□	□	□	□	□
(2) 朋友	□	□	□	□	□
(3) 家庭傭工	□	□	□	□	□
(4) 親戚	□	□	□	□	□
(5) 社會機構與專業人士	□	□	□	□	□

(b) 和過去（十年前）相比，現在你獲得上述***情緒性的支持***是：
(1) 比過去少； (2) 差不多或基本相同； (3) 比過去多。

20.(a) 在我有病，行動不便或是需要照顧的時候，他（她）能夠“照顧我，煮飯，或者做日常家務，或者照顧我的起居飲食，或者陪同出街或去醫院，日常家居維修等”

	①從無	②很少	③間中	④好多次	⑤幾乎每一次
(1) 主要家庭成員	□	□	□	□	□
(2) 朋友	□	□	□	□	□
(3) 家庭傭工	□	□	□	□	□
(4) 親戚	□	□	□	□	□
(5) 社會機構與專業人士	□	□	□	□	□

(b) 和過去（十年前）相比，現在你獲得上述的***照顧性支持***是：
(1) 比過去少； (2) 差不多或基本相同； (3) 比過去多。

21.(a) 在我感到孤單或悶的時候，他（她）能夠“陪伴我，陪我一起飲茶、或者出街，或者在家裏、社區中心陪我閑談，一起玩，一起參加活動等”

	①從無	②很少	③間中	④好多次	⑤幾乎每一次
(1) 主要家庭成員	□	□	□	□	□
(2) 朋友	□	□	□	□	□
(3) 家庭傭工	□	□	□	□	□
(4) 親戚	□	□	□	□	□
(5) 社會機構與專業人士	□	□	□	□	□

(b) 和過去（十年前）相比，現在你獲得上*陪伴上的支持*是：
(1) 比過去少； (2) 差不多或基本相同； (3) 比過去多。

22.(a) 在我遇到困難或問題，或是想獲得某方面資訊時，他（她）能夠"為我提供資料、建議，包括醫療、健康、保險、社會福利、康體活動、職業等"

	① 從無	② 很少	③ 間中	④ 好多次	⑤ 幾乎每一次
(1) 主要家庭成員	□	□	□	□	□
(2) 朋友	□	□	□	□	□
(3) 家庭傭工	□	□	□	□	□
(4) 親戚	□	□	□	□	□
(5) 社會機構與專業人士	□	□	□	□	□

(b) 和過去（十年前）相比，現在你獲得上述*信息上的支持*是：
(1) 比過去少； (2) 差不多或基本相同； (3) 比過去多

第六部分 黃金歲月

I 健康

回想過往一個月

23.身體上的痛楚有否妨礙你做必要的事情？
(1) 從來沒有； (2) 很少有； (3) 有時有； (4) 很多時有； (5) 時時有。

24.你是否需要藥物或治療才能維持身體功能去維持日常生活（普通的高血壓、糖尿病等慢性疾病不屬此類）？
(1) 從來沒有； (2) 很少有； (3) 有時有； (4) 很多時有； (5) 時時有。

25.你是否可以走動自如，四處走走？
(1) 完全不能夠； (2) 少許不能夠； (3) 某程度能夠；(4) 頗能夠； (5) 完全能夠。

26.總體來說，你現時的健康是：
(1) 極差； (2) 頗差； (3) 還可以； (4) 頗好； (5) 極好。

II 身體和認知功能

27.最近，你有足夠體力應付日常生活嗎？

(1) 很不足夠； (2) 不足夠； (3) 還可以；(4) 足夠； (5) 完全足夠。

28.最近，你睡得好嗎？
(1) 完全不好； (2) 不好； (3) 尚可； (4) 頗好； (5) 很好。

29. 在最近一個月，你有多少次無法記起熟悉的名稱或文字？
(1) 沒有； (2) 一兩次； (3) 三、四次； (4) 五、六次； (5) 七次或以上。

30.現時，你能否同時做兩項或以上的工作，例如能夠同一時間觀看電視，又與別人傾談與電視無關的事？
(1) 完全不能夠； (2) 需要留神才能夠； (3) 還可以； (4) 很能夠； (5) 完全能夠。

31.你現時的身體功能（包括體力、思考力、視力、聽和講的能力等）是否足夠使你過著獨立而無需他人照顧的生活？
(1) 很不足夠； (2) 不足夠； (3) 還可以； (4) 足夠； (5) 完全足夠。

III 關懷他人

32.你現時與主要家庭成員、親人交往或相處時，是否做過以下行為以關懷他們並幫助他們生活得更好？

	① 極多	② 頗多	③ 一般	④ 很少	⑤ 從未
(1) 願意承擔家庭事務（例如煮飯、清潔、洗衣及其他家務等）	☐	☐	☐	☐	☐
(2) 照顧家人（例如配偶、父母、兄弟姐妹、兒女或孫仔孫女）	☐	☐	☐	☐	☐
(3)為家人及親戚提供幫助（例如在經濟，情感上給予他們支持等）	☐	☐	☐	☐	☐

IV 老有所為

33. 在過去一年，你有沒有從事過下面的活動或者對社區、社會做出過貢獻？

	① 極多	② 頗多	③ 一般	④ 很少	⑤ 從未
(1) 義務工作	☐	☐	☐	☐	☐
(2) 受薪工作（例如全職和兼職）	☐	☐	☐	☐	☐
(3) 社交及休閒活動（例如社交聯誼活動，康樂活動，選舉投票，體育活動，藝術性活動，宗教活動，參與培訓課程等）	☐	☐	☐	☐	☐

V 生活滿意度

34. 你是否滿意目前的生活呢？請根據下面的描述，選擇一個最符合你的答案。

	① 非常反對	② 反對	③ 不同意不反對	④ 同意	⑤ 非常同意
(1) 大多數時候我感覺我的生活與我的理想相近	☐	☐	☐	☐	☐
(2) 我現在生活的條件是非常之好的	☐	☐	☐	☐	☐
(3) 我滿意我現在的生活	☐	☐	☐	☐	☐
(4) 到目前為止我已經獲得了生活中我想要的重要的東西	☐	☐	☐	☐	☐
(5) 如果我可以再活一次，我還是寧願不改變我的生活	☐	☐	☐	☐	☐

第七部分 應對能力

35. 請你回憶，在日常生活中如果遇到困難、壓力和不開心的事情時，你是否會採取以下的方式來渡過呢？

	① 從來沒有	② 有時有	③ 很多時候
(1) 工作或學習	☐	☐	☐
(2) 同人傾诉	☐	☐	☐
(3) 看到事物好的一面	☐	☐	☐
(4) 改變自己的想法	☐	☐	☐
(5) 不要將問題看得太嚴重	☐	☐	☐
(6) 堅持立場（認為自己的決定及做法是正確的）	☐	☐	☐
(7) 找出幾種解決方法	☐	☐	☐
(8) 向親戚朋友或舊同學尋求建議	☐	☐	☐
(9) 改變做法	☐	☐	☐
(10) 借鑒他人（參考其他人的做法）	☐	☐	☐
(11) 積極參加文化、運動或體育活動	☐	☐	☐
(12) 克制情感 （如衝動、生氣、悲傷、痛苦等）	☐	☐	☐
(13) 休息或休假	☐	☐	☐
(14) 吸煙、喝酒、服藥或者吃東西	☐	☐	☐

⒂ 等待 （等待事情的自然發展）	☐	☐	☐
⒃ 試圖忘記	☐	☐	☐
⒄ 依靠別人	☐	☐	☐
⒅ 接受現實	☐	☐	☐
⒆ 幻想奇跡 （希望有一天會有好的結果出現）	☐	☐	☐
⒇ 安慰自己	☐	☐	☐

第八部份 基本資料

36. 你的婚姻狀況是？
⑴ 從未結婚； ⑵ 分居； ⑶ 已婚； ⑷ 離婚； ⑸ 鰥寡； ⑹ 其它（請註明）_ 。

37.你現在的住屋類型：
⑴ 公共屋邨； ⑵ 長者住屋； ⑶ 自置物業（包括居屋）；
⑷ 租樓（全層）； ⑸ 租房； ⑹ 租床位；
⑺ 天台屋/木屋； ⑻ 政府與個人合資 ⑼其它住所（請注明）：________。

38.家庭成員數目及居住狀況。請問你 與誰一起居住？（住在同一個單位）

	數目	是否一起居住及數目
獨居	—	
配偶	—	
父親或母親（在世）		
兒子		
女兒		
兒媳		
女婿		
孫子或孫女		
兄弟或姐妹		
家庭傭工		
朋友	—	
親戚	—	

說明：如果是獨居老人，則在“獨居”一欄填 1；如果沒有一起居住，則填 0；必要時請標注數目，如與三個兒子同住， 則在“兒子” 一欄 填“3” 。

39.現時**你一個月**的收入有多少？ （注：包括薪水，退休金，租金，綜援，生果金，子女給的生活費和香港長者生活津貼等）（單位：HKD）
⑴ 沒有收入； ⑵ 2000 以下； ⑶ 2001-4000； ⑷ 4001-6000；
⑸ 6001-8000； ⑹ 8001-10000； ⑺ 10000 以上； ⑻ 不清楚。

40.**你的家庭**在過去一個月的總收入有多少？（包括主要家庭成員）（單位：HKD）
⑴ 5000 以下； ⑵ 5001-10000 ； ⑶ 10001-15000； ⑷ 15001-20000；
⑸ 20001-25000； ⑹ 25001-30000； ⑺ 30000 以上。

41. 總括來說，你覺得你的家庭的社會與經濟地位在香港屬於哪個水平？
⑴ 下層； ⑵ 中下層； ⑶ 中層； ⑷ 中上層； ⑸ 上層。

42. 你覺得現在是否有足夠金錢使用？
⑴ 十分夠； ⑵ 頗夠； ⑶ 剛剛夠； ⑷ 頗不夠； ⑸ 十分不夠。

43.在退休之前，你主要從事的職業是哪一種？
⑴ 一般工人（如服務業工人、售貨員、看更、地盤工人）；
⑵ 技工（如司機、水電師傅）
⑶ 管工 （如管理地盤工人的工頭）
⑷ 自雇或經營小生意（如的士車主、士多老闆）
⑸ 文職人員（如秘書、經紀、研究助理）
⑹ 技術人員（如電算師、社工、教師、護士）
⑺ 專業人員（如律師、工程師、教授、校長）
⑻ 行政人員（如經理、政務主任）
⑼ 初級軍警及紀律部隊人員
⑽ 公司老闆
⑾ 家庭主婦、主夫
⑿ 其他（請注明： ）。

44. 你有沒有患上以下的疾病（可選擇多項）：
⑴ 健康良好； ⑵ 血壓高/低； ⑶ 眼疾，如白內障； ⑷ 氣管疾病； ⑸ 心臟病； ⑹ 關節炎（腳痛、四肢痛、肩膀痛等）； ⑺ 糖尿病；

(8) 中風； (9) 腎病； (10) 帕金遜症； (11) 腦退化（老人癡呆） (12)痛風 (13) 其他（請注明：------------------）。

45. 在日常生活上你是否需要別人之協助？如果需要協助，誰是你的主要照顧者？
(1) 配偶； (2) 兒子； (3) 女兒； (4) 媳婦； (5) 女婿； (6) 孫子孫女；
(7) 家庭傭工； (8) 其他親戚； (9) 朋友； (10) 其他（請注明：--------------）

46. 你家中現時或最近是否有僱用家庭傭工？（如選 1，則無須回答 52-56 題）
(1) 從無； (2) 全職外傭； (3) 全職本地工人； (4) 兼職本地工人（鐘點工）。

47. 傭工主要為你提供哪些服務？ （可選擇多項）
(1) 基本家務（煮飯，清潔等）； (2) 個人照料（沖涼、剪指甲、穿衣等）；
(3) 協助醫療保健、物理治療、簡單護理等； (4) 照顧小朋友或家人；
(5) 陪同外出（包括看醫生，行街等）； (6) 聊天；
(7) 其他（請註明： ）。

48. 你對傭工提供的服務是否滿意？
(1) 非常不滿意； (2) 不滿意； (3) 一般； (4) 滿意； (5) 非常滿意。

49. 傭工每月需要支付多少錢？ 〔_______〕HKD。

50. 傭工的薪水由誰來支付？ （可選擇多項）
(1) 自己； (2) 配偶； (3) 兒子； (4) 媳婦； (5) 女兒； (6) 女婿；
(7) 孫子孫女； (8) 其他親戚； (9) 朋友； (10) 其他（請註明 ）。

51. 如果是自己與其他人分擔傭工的薪金，你自己要支付傭工多少薪水（每月） ？

52. 如果幾年之後，你身體狀況變差或是需要人照顧，你會願意找家庭傭工照料你嗎？（如目前已有家庭傭工，則無須回答此題）
(1) 願意 (2)不願意 (3)不清楚

53. 請談一下對雇傭家庭傭工作為照顧者的看法？優點與缺點都有哪些？--------------------------

----

54. 如果有需要，你是否願意參加**改善家居及社區照顧服務**，專業人員會為你提供個人照顧、復康運動以及基本護理等服務，以實現在家里或社區中安老的目的？
(1) 願意 (2)不願意 (3)不清楚

55. 如果你身體狀況越來越差，或者年紀越來越大，你是否願意入住老人院？
(1) 願意 (2)不願意 (3)不清楚

56. 你是否有宗教信仰？
(1) 無； (2) 基督教； (3) 天主教； (4) 佛教； (5) 拜祖先； (6) 伊斯蘭教；
(7)道教； (8) 其他（請註明__________________）。

訪問到此結束，多謝合作！

****

訪問員筆記

被訪者是否有嚴重的問題，需要聯繫社工或轉介？
訪問過程中有遇到什麼困難？
訪問的心得

Kolibri 360 GmbH
Ferdinand-Jühlke-Straße 7
99095 Erfurt, Deutschland
produktsicherheit@kolibri360.de